T0185717

Soft Error Reliability of VLSI Circuits

Behnam Ghavami • Mohsen Raji

Soft Error Reliability of VLSI Circuits

Analysis and Mitigation Techniques

 Springer

Behnam Ghavami
Shahid Bahonar University of Kerman
Kerman, Iran

Mohsen Raji
Shiraz University
Shiraz, Iran

ISBN 978-3-030-51612-3   ISBN 978-3-030-51610-9   (eBook)
https://doi.org/10.1007/978-3-030-51610-9

This Springer imprint is published by the registered company Springer Nature Switzerland AG
The registered company address is: Gewerbestrasse 11, 6330 Cham, Switzerland

"I want to thank my father and mother."
Thank you.
Dedicated to my family, Arezoo and Hiva.
Behnam
"I want to thank my father and mother."
Thank you.
Dedicated to my family, Zahra and Aala.
Mohsen

Preface

In recent decades, VLSI circuit designers have achieved high-speed circuits with fewer power requirements by continuously reducing the transistor dimensions. Besides these benefits, the digital circuits have become more sensitive against environmental faults due to several factors such as decreasing the transistor size and the supply voltage. Hence, the reliability of the VLSI circuits has become an important issue in recent years. Among these noise sources, radiation-induced soft errors in commercial nanometer CMOS technologies have become a growing concern. Nanometer integrated circuits are getting increasingly vulnerable to soft errors making the soft error rate (SER) estimation and optimization an important challenge. Efficient and accurate SER estimation and optimization of the large-scale combinational circuits help the designers to achieve more reliable integrated circuits.

This book will investigate emerging trends in the design of today's reliable electronic systems which are applicable to safety-critical applications like automotive or healthcare electronic systems. The book will introduce software tools for analysis and mitigation of soft errors in electronic systems. The emphasis is on modeling approaches and algorithms for analysis and mitigation of soft error issue in nanoscale CMOS digital circuits and techniques that are the cornerstone of Computer Aided Design (CAD) of a reliable VLSI circuit.

Kerman, Iran
Shiraz, Iran

Behnam Ghavami
Mohsen Raji

Acknowledgments

After one decade of researching and teaching in the field of fault-tolerant computing systems, we decided to write this book. The groups of students from Shahid Bahonar University of Kerman who developed the algorithms and software packages are important contributors of this work.

We would like to thank Professor Hossein Pedram for keeping us engaged and learning. Our special thanks go to our graduated students Amin Sabet, Mohammad Reza Rohanipour, and Mohammad Eslami who developed the materials for chapters and helped with the preparation of the book. All this would not have been possible without their valuable support.

Writing this book is harder than we thought and more rewarding than we could have ever imagined. None of this would have been possible without our families.

Behnam and Mohsen
April 2020

Contents

Chapter 1
Introduction: Soft Error Modeling

In recent decades, VLSI circuit designers have achieved high-speed circuits with fewer power requirements by continuously reducing the sizes of transistors. Besides these benefits, aggressive technology scaling has led to serious reliability challenges for nanoscale digital circuits. Various sources that may affect the circuit reliability can be divided into two groups: permanent faults and transient faults. Permanent faults are the ones which lead to constant malfunctioning of the circuit. Examples of sources that lead to permanent faults are hot-carrier injection (HCI) and negative-bias temperature instability (NBTI) in transistors and electrical migration between connections. In addition to permanent fault, transient faults can also have a negative effect on correct functionality of the circuit. Transient faults occur in certain environmental conditions and are present in the system for a short time and may cause an error in the circuit or disappear without causing a problem for the circuit. Among various sources resulting in transient fault, collisions of high-energy particles emitted by cosmic rays and emission of alpha particles from the circuit package to sensitive areas of a semiconductor have had the greatest negative impact on the reliability of digital circuits [1].

The transient faults caused by a high-energy particle strike in the combinational circuits and sequential elements (i.e., memory cells, latches, and flip-flops) are called single-event transient (SET) and single-event upset (SEU), respectively [1]. In recent years, SET has been one of the most important sources that could affect the reliability of the combinational circuits. Such a bit-flip of a storage node is called a single-event upset (SEU). This chapter provides introductions to soft error issue and soft error modeling procedure.

This chapter is organized as follows. Section 1.1 explains the impacts of particle strike on semiconductor devices. In Sect. 1.2, we explain the modeling approach of single-event transient and single-event upset and in Sect. 1.3, we describe the impacts of particle hit on combinational circuits.

© Springer Nature Switzerland AG 2021
B. Ghavami, M. Raji, *Soft Error Reliability of VLSI Circuits*,
https://doi.org/10.1007/978-3-030-51610-9_1

1.1 Impacts of Particle Strike on Semiconductor Devices

When high-energy particles strike into a sensitive part of a semiconductor device (Fig. 1.1a), a dense channel of electrons and holes (called funnels) are formed when passing through a P and N junction (Fig. 1.1b). Then, a number of these electrons and holes are recombined together, which is called the charge collection phenomenon (Fig. 1.1c). This process leads to the production of a current with a very short duration at the collision site.

The maximum amount of the induced current and the width of the generated SET depend on the process of collecting the charge after the production of the electron-hole channel [2]. Factors such as channel density and charge collection velocity determine the shape of the produced SET. The total charge collection process depends on the concentration of the substrate; that is, the charge collection will continue until the density of the generated carriers becomes equal to the impurity of the substrate. If the impurity of the substrate is low, more carriers will be generated leading to higher induced current.

The height and width of the induced current pulse depend on the amount of energy of the striking particle. In other words, it depends on the amount of charge induced by the high-energy particle. The linear energy transfer (LET) is a parameter used to show the relationship between the particle energy and the amount of charge generated in a particular type of material. One LET is the amount of energy that is lost by a particle during its motion in a unit of material length. The unit of LET is MeV cm^2/mg.

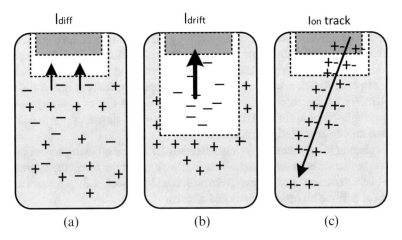

Fig. 1.1 Overall mechanism of a particle strike into a P and N junction: (**a**) particle striking moment, (**b**) charge collection, (**c**) redistribution

1.2 Single-Event Transient Modeling

The process of striking the particles to the surface of the circuit and generation of a single-event transient is illustrated by a current model [3]. After striking the particle on the silicon, the charge collection phenomenon occurs leading to appearance of a transient current in the strike point. The generated current from the particle is approximately modeled by an exponential equation as

$$i_{coll}(t) = \frac{Q_{coll}}{\tau_{fall} - \tau_{rise}} \left\{ \exp\left(\frac{-t}{\tau_{fall}}\right) - \exp\left(\frac{-t}{\tau_{rise}}\right) \right\} \tag{1.1}$$

where Q_{coll} is the amount of the induced charge in the semiconductor element caused by the particle strike. τ_{rise} and τ_{fall} are, respectively, the rising time and the delay of the current pulse which are technology dependent. The slope and the width of the generated pulse depend on the circuit and the environmental parameters such as the angle of the strike and the amount of impurity in the materials, which can be modeled by changing the timing parameters of Eq. (1.1).

The induced charge caused by a particle hit and, consequently, the generated current may provide the necessary charge to change the stored value in a memory cell. In this case, a single-event upset (SEU) or a soft error will occur. Of course, this particle can hit a transistor in the combinational part of the circuit leading to another issue as described in the following.

1.3 Impact of a Particle Hit on Combinational Circuits

Figure 1.2 shows the general process of a particle hit on a sensitive area of an inverter gate. The particle hit on a PMOS transistor in an OFF state generates a current pulse in this transistor. The produced current charges the load capacitor (C_L) in the output node of the inverter gate and then the load stored in the capacitor will be discharged from the capacitor through the NMOS in the ON state. This process leads to a transient voltage pulse generation at the output of the circuit. This voltage pulse is called a single-event transient or briefly SET.

The particle hits on a transistor in a circuit may result in a SET at the output of the circuit gate. In order to cause a soft error in the circuit, the generated SET has to be logically propagated along the combinational circuit and eventually reach the input of a memory element such as latch or flip-flop and be stored in it. During propagation, the logical, electrical, and timing characteristics of the SET may be affected preventing the SET to be changed into a soft error. These properties, which are called masking mechanisms, are decreased in nanoscale technologies. The details of masking mechanisms are described in the following subsections.

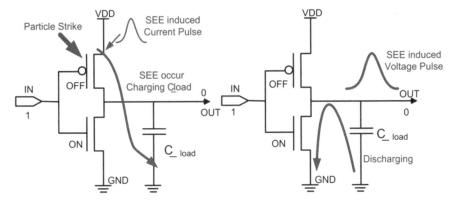

Fig. 1.2 The process of particle strike to an inverter gate and the induced current pulse in its output [4]

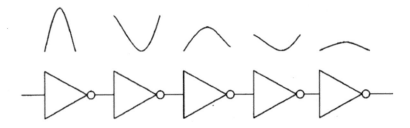

Fig. 1.3 An illustrative example of electrical masking effect

1.3.1 Electrical Masking

During transient pulse propagation through a combinational logic circuit, the height and width of the SET may be attenuated as it passes through the input to the output of each gate. If the SET height decreases such that it does not actually show a logical change in its value, it will not be propagated through the gate. Moreover, if the pulse width decreases such that it does not meet the time requirements of a memory element, it will not be stored by the memory element. In these cases, it is called that the SET is electrically masked [5]. The amount of electrical masking that occurs at each gate is a function of the gate delay as well as the height and width of the SET [6, 7]. Figure 1.3 shows an example of a SET being attenuated when passing through a chain of logical gates. The characteristics of the SET decrease with successive passes through the gates in the logic circuit.

The overall procedure of the technology improvement and decrease in the pipeline levels for reducing the clock cycle timing decrease the electrical timing masking effect, because the circuit depth in each pipeline level will be decreased. Furthermore, shrinking in size due to consecutive improvement of technology generations affects the electrical masking effect from two points of view. First, smaller size leads to decrease in the amount of the critical charge in the circuit which means that the different parameters of the SET are being reduced. Also, the individual attenuation of the SET in each gate is decreased since the propagating delay is reduced due to the technology scaling. So, the electrical masking effect is decreased and the circuits become more vulnerable against the soft error.

1.3.2 Logical Masking

When a particle strikes the combinational parts of the circuit, the generated transient pulse may cause a change in latching elements only if there is a sensitized path between the strike point and the latching element. The existence of the sensitized path is based on the primary inputs of the circuit. For example, in the circuit shown in Fig. 1.4, the transient pulse generated in the output of G_a is not propagated to the output since the other input of the G_b has the logical value of 0, and the output remains the logical value of 0. So, the SET is masked in this part of the circuit and does not propagate through the circuit. This process is called logical masking [5]. The amount of logical masking in a circuit depends on the implementing function of the circuit and is independent from all technology parameters, and as a result technology improvement does not affect the logical masking.

1.3.3 Timing Masking

Even if a SET is not masked by electrical and logical masking mechanisms, it still needs to reach the input of the latching element in a specific time interval. This time interval in which the latching element stores a new value is known as latching window. It is calculated by summation of the setup and hold time [5]. The pulses reaching to the inputs of the latching elements out of this latching window are masked by timing masking effect. Figure 1.5 illustrates the different timing masking mechanisms. In this figure, it is assumed that the sequential element of the flip-flop is

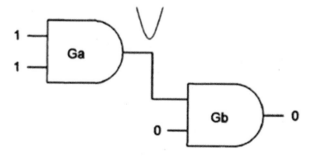

Fig. 1.4 An example of logical masking

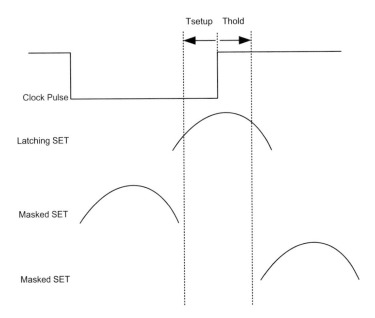

Fig. 1.5 An example of timing masking

triggered by the rising edge of the clock. The waveform on the top of the figure indicates the clock cycle signal and the dotted lines show the latching window. As shown in this figure, a transient pulse changes the value of the flip-flop only if it occurs inside the latching window of the flip-flop.

Continuous technology scaling reduces the timing masking. Decrease in the critical charge means that the propagating transient pulses have larger widths and decrease in gate delays means both the clock pulse length and the latching window are decreased. The combination of these effects increases the probability of reaching a transient pulse to the input of a flip-flop, which means that the timing masking occurs less.

References

1. R. C. Baumann, "Soft Errors in Advanced Semiconductor Devices—Part I: The Three Radiation Sources," IEEE Transactions on Device Material Reliability (TDMR), Vol. 1, No. 1, pp. 17–22, 2001.
2. M. Zhang, N. R. Shanbhag, "A Soft Error Rate Analysis (SERA) Methodology," in Proceedings of the IEEE International Conference on Computer Aided Design (ICCAD), pp. 111–118, 2004.
3. G. R. Srinivasan, P. C. Murley, H. K. Tang, "Accurate, Predictive Modeling of Soft Error Rate due to Cosmic Rays and Chip Alpha Radiation," in Proceedings of IEEE International Reliability Physic Symposium (IRPS), pp. 12–16, 1994.

4. M. FAZELI, S. G. MIREMADI, H. ASADI AND S. N. AHMADIAN, "A FAST AND ACCURATE MULTI-CYCLE SOFT ERROR RATE ESTIMATION APPROACH TO RESILIENT EMBEDDED SYSTEMS DESIGN," IN 2010 IEEE/IFIP INTERNATIONAL CONFERENCE ON DEPENDABLE SYSTEMS & NETWORKS (DSN), CHICAGO, IL, 2010, PP. 131–140.
5. P. SHIVAKUMAR, M. KISTLER, S. W. KECKLER, D. BURGER, L. ALVISI, "MODELING THE EFFECT OF TECHNOLOGY TRENDS ON SOFT-ERROR RATE OF COMBINATIONAL LOGIC," IN PROCEEDINGS OF INTERNATIONAL CONFERENCE ON DEPENDABLE SYSTEMS AND NETWORKS (DSN), PP. 389–398, 2002.
6. M. OMANA, G. PAPASSO, D. ROSSI, C. METRA, "A MODEL FOR TRANSIENT FAULT PROPAGATION IN COMBINATIONAL LOGIC," IN PROCEEDINGS OF IEEE INTERNATIONAL ON-LINE TEST SYMPOSIUM (IOLTS), PP. 111–115, 2003.
7. R. RAJARAMAN, J. S. KIM, V. NARAYANAN, Y. XIE, M. J. IRWIN, "SEAT-LA: A SOFT ERROR ANALYSIS TOOL FOR COMBINATIONAL LOGIC," IN PROCEEDINGS OF THE INTERNATIONAL CONFERENCE ON VLSI DESIGN, PP. 159–165, 2006.

Chapter 2
Soft Error Rate Estimation of VLSI Circuits

2.1 Introduction

Soft error rate (SER) is a measurement metric which is used to evaluate the sensitivity of a digital circuit to SETs. Efficient and accurate SER estimation of the combinational circuits helps the designers to achieve more reliable integrated circuits. Traditional SER estimation techniques can be categorized into two major groups: dynamic and static approaches. Dynamic approaches are based on fault injection and logic simulations. Although dynamic methods provide highly accurate SER estimation results, they are drastically time consuming and are intractable for large circuits [1–5]. On the other hand, static approaches take advantage of mathematical modeling or special data structures to model the initial parameters of SETs and also the SET propagation model in the SER estimation process.

In this chapter, we present a method for soft error rate estimation of the combinational circuits based on vulnerability evaluation of the circuit gates to soft errors [6]. To evaluate the vulnerability of the circuit to SETs, we introduce a concept called probabilistic vulnerability window (PVW) which considers the impacts of triple masking factors (i.e., logical, electrical, and timing). In other words, the parameters of PVW can be considered as a unified inference of various constraints necessary for an SET to cause soft errors in the circuit when propagating through the gates on its paths to the primary outputs. Using a computation model, PVWs for each circuit gate output are calculated based on a level-order backward circuit traversing and meanwhile the SER of the circuit is gradually estimated considering a mathematical formulation of SER considering the computed PVW parameters. In order to validate the presented approach, we compare the obtained results with Monte Carlo-based fault injection (MCFI) simulations. This comparison indicates that the accuracy of the presented method is in average 2% of the results obtained by MCFI with four orders of magnitude speedup.

This chapter is organized as follows. Section 2.2 formulates the SER estimation of a given combinational circuit. In Sect. 2.3, we describe the details of PVW and in

© Springer Nature Switzerland AG 2021
B. Ghavami, M. Raji, *Soft Error Reliability of VLSI Circuits*,
https://doi.org/10.1007/978-3-030-51610-9_2

Sect. 2.4, we present the approach for error probability computation. Section 2.5 presents the SER estimation algorithms. In Sect. 2.6, we report the experimental results for a set of common benchmarks. Finally, Sect. 2.7 concludes this chapter.

2.2 Soft Error Rate Estimation Formulation

Figure 2.1 shows a general view of a combinational circuit including the combinational logic gates, as well as primary inputs (PI) and primary outputs (PO). A SET may appear in the output of the internal logic gates in addition to the interconnection between PIs and internal gates. We call each of these places in the circuit as a circuit node (CN). Circuit node set (CNS) is defined as the set which includes all CNs. The SET originated at any CNs of CNS may be latched in each element considered in POs.

Soft error rate of such a combinational circuit (SER_c) can be calculated as follows [4]:

$$\text{SER}_c = R_{\text{PH}} \cdot \alpha \cdot A_c \cdot P\left(\text{SE}_c\right) \tag{2.1}$$

where R_{PH} is the particle hit rate per unit of area, α is the fraction of effective particle hits (those resulting in charge generation), A_c is the total silicon area of the circuit, and $P(\text{SE}_c)$ is the probability of occurrence of a soft error in the combinational circuit conditioned on an effective particle hit. Details about the value of $R_{\text{PH}} \bullet \alpha$ are well provided in [4, 7]. Hence, we focus on estimating the probability that a SET at a given gate G of the circuit will result in a soft error of at least one of the POs.

Based on [6], the probability of SE_c can be calculated as

$$P\left(\text{SE}_c\right) = \frac{1}{N_{\text{CNS}}} \sum_{i=1}^{N_{\text{CNS}}} \int_{w_{\text{min}}}^{w_{\text{max}}} P\left(E_{i,w}\right) \cdot f\left(w\right) \cdot dw \tag{2.2}$$

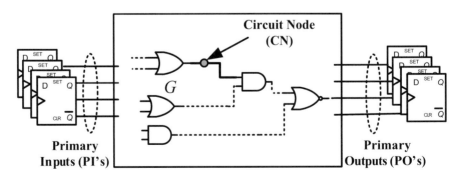

Fig. 2.1 A general view of a combinational circuit

where:

- N_{CNS} is the number of circuit nodes (i.e., circuit gate outputs), and (w_{min}, w_{max}) are the (minimum, maximum) width of SETs in the circuit nodes.
- $E_{i, w}$ is the event of any PO failing (experience of a soft error) given that the SET occurs at the output of the CN_i with the initial width of w and $P(E_{i, w})$ is the probability of $E_{i, w}$.
- $f(w)$ is the probability density function of random variable $\mathcal{W} \in (w_{min}, w_{max})$.

The type of function $f(w)$ depends on different factors such as the particles' energy and the critical charge for circuit transistors [8, 9]. There have been several works that present different ways to determine $f(w)$ [4, 10]. Hence, in order to calculate the probability of SE_c, we focus on calculating the probability of $E_{i, w}$ (which will be noted as E_i hereinafter for simplicity).

In its turn, E_i can be defined formally as

$$E_i = \bigcup_{i=1}^{N_{PO}} E_i^j \tag{2.3}$$

where N_{PO} is the number of POs and E_i^j is the event of a soft error observed in PO j due to the SET that occurred in CN_i with the initial width of w.

Using the addition law of probability [6], Eq. (2.3) can be written as

$$P(E_i) = P\left(\bigcup_{j=1}^{N_{PO}} E_i^j\right)$$

$$= \sum_{j_1=1}^{N_{OP}} P\left(E_i^{j_1}\right) + \sum_{k=2}^{N_{PO}} (-1)^{k+1} \sum_{j_1=1}^{N_{PO}} \sum_{j_2=j_1+1}^{N_{PO}} \cdots \sum_{j_k=j_{k-1}+1}^{N_{PO}} P\left(\bigcap_{n=1}^{k} E_i^{j_n}\right) \tag{2.4}$$

where $P\left(\bigcap_{n=1}^{k} E_i^{j_n}\right)$ refers to the probability of the event that the occurred SET in the CN_i will result in all PO j_1 to j_k.

$P\left(\bigcap_{n=1}^{k} E_i^{j_n}\right)$ is equal to zero when there is no chain of connected logic gates or *logical path* from CN_i to all PO j_1 to j_k; otherwise it is not equal to zero. In order to reduce the complexity of the estimation process, we assume that different primary outputs of the circuit are independent [3, 11, 12]. So, the overall error probability in (2.4) can be calculated as

$$P(E_i) = \sum_{j_1=1}^{N_{OP}} P\left(E_i^{j_1}\right)$$

$$+ \sum_{k=2}^{N_{PO}} (-1)^{k+1} \sum_{j_1=1}^{N_{PO}} \sum_{j_2=j_1+1}^{N_{PO}} \cdots \sum_{j_k=j_{k-1}+1}^{N_{PO}} \left(1 - \prod_{n=1}^{k} \left(1 - P\left(E_i^{j_n}\right)\right)\right) \tag{2.5}$$

Hence, in order to evaluate the SER of the circuit, it is enough to calculate the error probability of each CN_i regarding each PO (i.e., $P\left(E_i^{j_n}\right)$). In the following, we firstly introduce the details of PVW and then, in Sect. 2.4, we describe how to use PVW to determine $P\left(E_i^{j_n}\right)$.

2.3 Probabilistic Vulnerability Window

In this section, we give a conceptual description of PVW and after that, we present a computation model which shows how PVWs of different circuit nodes are determined.

2.3.1 PVW Definition

Considering three masking factors (logical, electrical, and timing), we introduce PVW which is an inference of the requirements that should be met by an SET in a given circuit node in order to be latched as a soft error. A PVW of a given circuit node (CN_i) consists of four different parameters, i.e., (OID_w, ST_w, $W_w^{0(1)}$, P_{sen}), which are characterized as follows:

- *Output ID* (OID_w): The identification number of the memory element in which the soft errors are latched due to the SETs that occur in a given circuit node
- *Starting time* (ST_w): This parameter shows the constraint on the starting time from which the assumed SET in the given circuit node should begin in order to be latched as a soft error.
- *Width pairs* (W_w^0 and W_w^1): This pair of parameters represents the minimum width of a given SET (with value changes in 0-1-0 and 1-0-1) which is required for it to be latched as a soft error.
- *Sensitization probability* (P_{sen}): A probabilistic value which is associated with PVW. This value shows the probability of the event that a SET originated in the given CN_i finds a logical path to the corresponding output.

Parameters of PVW show the triple constraints which are necessary to be satisfied by a SET to be latched as a soft error. W_w^0 and W_w^1 represent the electrical masking factor as they show the minimum required value for a SET width to be latched as a soft error. Regarding the timing masking factor, ST_w shows the minimum time interval within which the SET should occur so that it reaches the memory element by the required latching timing window. P_{sen} is related to the logical masking factor as it shows the probability that a SET is not masked due to the logical values of the other inputs of the gate propagating it through the logical path from the circuit gate output to PO. Hence, PVW considers the vulnerability of the circuit nodes to SET and the resulting soft errors.

2.3.2   PVW Computation Model

In order to calculate the PVWs for all circuit nodes, we firstly initiate the parameters of the PVW for each PO and then traverse the circuit graph backward and compute PVW parameters for all other circuit nodes.

Without loss of generality, we assume that, in PO j, there is a flip-flop with setup time (t_s) and hold time (t_h) and the clock frequency is equal to T_{clk} (Fig. 2.2a). We take $[0, T_{clk}]$ as the interval of observation as we are calculating the occurrence of a soft error due to the propagation of a SET in the time interval between two active clock edges (i.e., rising/falling). Considering the timing masking factor, a signal needs to be stable for an interval from the setup time (t_s) before the rising edge of the clock until the hold time (t_h) after the rising edge of the clock in order to be latched in the FF. Considering this fact, Table 2.1 indicates the initiation value assigned to the parameters of PVW related to PO j and also a justification for our value assignments.

For the other nodes except POs, we use a computation model which helps us to determine the parameters of PVWs at those nodes. In Fig. 2.2b, we show an example of a circuit in which the output of logical gate related to CN_i is connected to the input in_{k_i} of the logical gate with propagation delay t_p related to the circuit node CN_{i^*}. Given that the parameters of $PVW_{i^*}^j$ are determined, Table 2.2 shows how to calculate the parameter for PVW_i^j.

2.4   Error Probability Computation

In this section, we describe how to use PVWs to calculate the SER of a circuit by computing the error probability of each CN_i regarding each PO (i.e., $P\left(E_i^{j_n}\right)$).

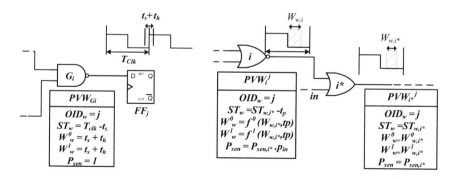

(a) PVW initialization   (b) General View of PVW Computation

Fig. 2.2 A sample of PVW parameter computation. (**a**) PVW initialization. (**b**) General view of PVW computation

Table 2.1 PVW parameter initiation and justifications

Parameter	Initiation value	Justification
OID_w	j	The circuit node is connected to the PO j
ST_w	$T_{clk} - t_s$	The SET originated at PO j should start (reaches to the switching threshold of the FF in PO j) before the time $T_{clk} - t_s$; otherwise it will not be considered as a valid value by FF in PO j and will be masked through timing masking factor [10]
$\left(W_w^0, W_w^1\right)$	$t_s + t_h$	The SET originated at the input of PO j should continue onto $T_{clk} + t_h$ which means that its width should be more than $t_s + t_h$; otherwise it is assumed to be masked through timing masking factor [10]. Since the type of SET (i.e., 0-1-0 or 1-0-1) makes no difference in the latching process, the initial value is equal for both elements of this pair
P_{sen}	1	There is no gate in the path of occurred SET to the PO; there is certainly a synthesized path to PO

Considering the definition of $E_i^{j_n}$ and logical, electrical, and timing masking factors, we define three events of $E_{i,l}^{j_n} / E_{i,e}^{j_n} / E_{i,t}^{j_n}$ as follows: the SET occurred in CN_i to be latched as a soft error in PO j despite the impacts of *logical/electrical/timing* masking factors. These events can be formally defined based on the parameters of PVWs.

In order to be latched at the PO j_n, the SET occurring in the CN_i should satisfy the setup time and hold time conditions as it has been assumed that there is a FF in PO j_n. Considering the definition of PVW, this condition implies the necessary conditions for the parameters of the SET at CN_i as follows:

- The starting time of SET should be less than the parameter ST_w.
- The width of the SET at CN_i has to be larger than the $\left(W_w^0, W_w^1\right)$ of $PVW_i^{j_n}$.

So, for the starting time (t_{s,SET_i}) and ending time of transition of SET with width W_{SET_i} in CN_i, we have

$$t_{s,SET_i} < ST_w \tag{2.6}$$

$$t_{s,SET_i} + W_{SET_i} > ST_w + W_w \tag{2.7}$$

where W_w represents both W_w^0 and W_w^1 parameters. Thus, we can express the condition by

$$t_{s,SET_i} \in \left[ST_w + W_w - W_{SET_i}, ST_w \right] \tag{2.8}$$

In addition, for the width of the SET, we have

$$W_w < W_{SET_i} < \tau \tag{2.9}$$

Table 2.2 PVW parameter computation model

Parameter	Propagated value	Justification
$PVW_i^j \cdot OID_w$	$PVW_{i^*}^j \cdot OID_w$	This parameter is not changed.
$PVW_i^j \cdot ST_w$	$PVW_{i^*}^j \cdot ST_w - t_p$	As the SET in i^* has to propagate through a gate g with delay propagation equal to t_p, it should start in a time t_p before the time it should start in j.
$PVW_i^j \cdot \left(W_w^0, W_w^1\right)$	$f^x\left(PVW_{i^*}^j \cdot W_w^x\right)$: if $\left(PVW_{i^*}^j \cdot W_w^x > t_{rf} \cdot 1.25 + (-1)^x \cdot \left\|t_{plh} - t_{phl}\right\|\right)$: $PVW_i^j \cdot W_w^x + (-1)^x \cdot \left\|t_{phl} - t_{plh}\right\|$ $f^x\left(PVW_{i^*}^j \cdot W_w^x\right)$: if $\left(PVW_{i^*}^j \cdot W_w^x < t_{rf} \cdot 1.25 + (-1)^x \cdot \left\|t_{plh} - t_{phl}\right\|\right)$: $$\frac{(t_{rf} \cdot 1.25)\left(PVW_{i^*}^j \cdot W_w^x + t_{phl} + t_{plh}\right)}{(t_{rf} \cdot 1.25) + t_{phl} + t_{plh}}$$	This model is based on the glitch attenuation model used in [1]. x can be either 1 for PVW related to 0-1-0 SETs or 0 for PVW related to 1-0-1 SETs. If the gate i^* has a non-inverting Boolean logic (like AND, OR), the computed result of f^0 and f will be assigned to $PVW_i^j \cdot W_w^0$ or $PVW_i^j \cdot W_w^1$, respectively. Otherwise, if i^* is a gate with an inverting Boolean logic (like NOT, NAND, NOR), the result of computation of f^0 and f will be assigned to $PVW_i^j \cdot W_w^1$ or $PVW_i^j \cdot W_w^0$, respectively. t_{rf} refers to the rising/falling time of the gate i^* which is obtained by a pre-characterization process. For a 0-1-0 SET (with $PVW_{i^*}^j \cdot W_w^0$) t_{rf} equals the rising time and for a 1-0-1 SET (with $PVW_{i^*}^j \cdot W_w^1$), t_{rf} is equal to the falling time of i^*.
$PVW_i^j \cdot P_{sen}$	$PVW_{i^*}^j \cdot P_{sen} \cdot \Pr\left(\frac{\partial f_i}{\partial in_{k_i}}\right)$	This approach is similar to the gate error model used in [13] using Boolean difference calculus. $\Pr(\bullet)$ represents the signal probability function and returns the probability of its Boolean argument to be "1" and $\frac{\partial f_i}{\partial in_{k_i}}$ is the partial Boolean difference of the Boolean function f of CN, with respect to a single variable, in_{k_i}, returning the condition (on the rest of the variables) under which function f is sensitive to the input variable in_{k_i}.

where τ is the maximum possible width for SETs in a specific technology node [8, 9]. Please note that W_{SET} and τ depend on different parameters especially the operating environment. In space applications, τ will be higher than in standard commercial applications because of the presence of highly energetic particles.

More formally, we redefine events $E_{i,e}^{j_n}$ and $E_{i,t}^{j_n}$ in terms of (2.8) and (2.9) as

$$E_{i,e}^{j_n} : W_w < W_{SET_i} < \tau \tag{2.10}$$

$$E_{i,t}^{j_n} : t_{s,SET_i} \in \left[ST_w + W_w - W_{SET},ST_w \right] \tag{2.11}$$

Considering the effects of all three masking factors, the events $E_{i,l}^{j_n}$, $E_{i,e}^{j_n}$, and $E_{i,t}^{j_n}$ should occur simultaneously. So, the probability of $E_i^{j_n}$ can be calculated as the probability of the intersection of these events, i.e.,

$$P\left(E_i^{j_n}\right) = P\left(E_{i,l}^{j_n} \cap E_{i,e}^{j_n} \cap E_{i,t}^{j_n}\right) \tag{2.12}$$

However, as the logical masking depends on the Boolean functions implemented by the logical gates and the electrical and timing masking factors are dependent on the propagation delay of the gates, we can conclude that $E_{i,l}^{j_n}$ and $E_{i,e}^{j_n} \cap E_{i,t}^{j_n}$ are independent events. Hence, we have

$$P\left(E_{i,l}^{j_n} \cap \left(E_{i,e}^{j_n} \cap E_{i,t}^{j_n}\right)\right) = P\left(E_{i,l}^{j_n}\right) \cdot P\left(E_{i,e}^{j_n} \cap E_{i,t}^{j_n}\right) \tag{2.13}$$

Considering the definition of parameter P_{sen} of PVW, the probability of $E_{i,l}^{j_n}$ is equal to parameter P_{sen} of the corresponding PVW which has been computed in CN_i, i.e.,

$$P\left(E_{i,l}^{j_n}\right) = P_{sen} \tag{2.14}$$

The probability of the intersection of $E_{i,e}^{j_n}$ and $E_{i,t}^{j_n}$ can be written as

$$
\begin{aligned}
P\left(E_{i,e}^{j_n} \cap E_{i,t}^{j_n}\right) &= P\left(t_{s,SET_i} \in \left[ST_w + W_w - W_{SET_i},ST_w \right] \cap W_{SET_i} > W_w \right) \\
&= P\left(t_{s,SET_i} \in \left[ST_w + W_w - W_{SET_i},ST_w \right] \cap \left(\bigcup_k W_{SET_i} = W_{SET_i,k} \right)\right) \\
&= \sum_k \left(P\left(t_{s,SET_i} \in \left[ST_w + W_w - W_{SET_i},ST_w \right] | W_{SET_i} = W_{SET_i,k}\right) \cdot P\left(W_{SET_i} = W_{SET_i,k}\right)\right)
\end{aligned} \tag{2.15}
$$

where $\{ W_{SET_i,k} \}$ is the set of possible SET widths which may occur in CN_i.

Similar to [10–12], we assume that t_{s,SET_i} is uniformly distributed in the interval $(T_1, T_1 + T_{clk} - W_{SET_i,k})$ (T_1 equals the maximum delay from PIs to CN_i). Thus, in the

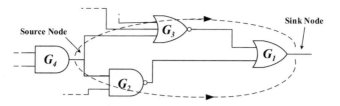

Fig. 2.3 An example of a reconvergent path in a circuit

worst case when, for a given SET duration $W_{\text{SET},k}$, the interval $\left[\text{ST}_w + W_w - W_{\text{SET}_i}, \text{ST}_w \right]$ lies inside it we have

$$P\left(t_{s,\text{SET}_i} \in \left[\text{ST}_w + W_w - W_{\text{SET}_i}, \text{ST}_w \right] | W_{\text{SET}_i} = W_{\text{SET}_i,k} \right)$$
$$= \frac{W_{\text{SET}_i,k} - W_w}{T_{\text{clk}} - W_{\text{SET}_i,k}}, \quad W_w < W_{\text{SET}_i} < \tau \tag{2.16}$$

Considering (2.12)–(2.16), the value of $P\left(E_i^{j_n} \right)$ can be evaluated which is the main part of SER estimation of a combinational circuit as mentioned in Sect. 2.2.

It is notable that this equation calculates a situation in which only one PVW has been computed to CN_i. However, there is a different situation in which more than one PVW may be propagated to the circuit node. This situation occurs when there is more than one path from POs to CN_i. In other words, there is a reconvergent path in the circuit. In the following, we extend the proposed model to address the reconvergent path case.

2.4.1 Handling Reconvergent Paths

Figure 2.3 shows an example of a circuit with a reconvergent path. We call a circuit node with a reconvergent fan-out as the *source node* and the circuit node in which the logical paths will reach together is called as the *sink node*.

Considering Fig. 2.3, the PVWs of the corresponding circuit node of G_5 are calculated based on the PVWs of the circuit node related to G_2, G_3, and G_4. Hence, there are at least two PVWs for CNs related to G_5. So, the number of PVWs is more than one in the source nodes of reconvergent paths. We gather all PVWs calculated for a given source node CN_i with the same OID j_n in a set called $\text{PVW}_i^{j_n}$, i.e.,

$$\text{PVW}_i^{j_n} = \left\{ \text{PVW}_{i,k}^{j_n} | k = 1 \ldots N_{\text{RF}} \right\} \tag{2.17}$$

where N_{RF} is the number of PVWs in set $\text{PVW}_i^{j_n}$.

In order to address the reconvergent path case, regarding the event $E_i^{j_n}$ in source node CN_i, we have N_{RF} events which may result in a soft error in the circuit, i.e., an event leading to error by propagating SET through k paths ($k = 1$ to N_{RF}) from CN_i to PO j_n. We show these events by $E_{i,k}^{j_n}$. The error probability of CN_i related to PO j_n can be computed as

$$P\left(E_i^{j_n}\right) = P\left(\bigcup_{k=1}^{N_{RF}} E_{i,k}^{j_n}\right) = \sum_{k=1}^{N_{RF}} P\left(E_{i,k}^{j_n}\right)$$
$$+ \sum_{m=2}^{N_{RF}} (-1)^{m+1} \sum_{k_1=1}^{N_{RF}} \sum_{k_2=k_1+1}^{N_{RF}} \cdots \sum_{k_m=k_{m-1}+1}^{N_{RF}} P\left(\bigcap_{l=1}^{m} E_{i,k_l}\right) \tag{2.18}$$

where the value of $P\left(E_{i,k}^{j_n}\right)$ can be computed based on (2.12)–(2.16) using the parameters of individual PVWs in $PVW_i^{j_n}$.

The value of $P\left(\bigcap_{l=1}^{m} E_{i,k_l}\right)$ can be determined using the similar approach used in this section. For the general case with m PVWs, $P\left(\bigcap_{l=1}^{m} E_{i,k_l}\right)$ can be calculated as

$$P\left(\bigcap_{l=1}^{m} E_{i,k_l}\right) = \left(\prod_{k=1}^{m} P\left(E_{i,l}^{k}\right)\right)$$
$$\times \sum_{k}\left(\frac{W_{SET_i,k} - \left(PVW_i \cdot ET_w - PVW_i \cdot BT_w\right)}{T_{clk} - W_{SET_i,k}} \cdot P\left(W_{SET_i} = W_{SET_i,k}\right)\right) \tag{2.19}$$

such that $\left(PVW_i \cdot ET_w - PVW_i \cdot BT_w\right) < W_{SET_i,k} < \tau$.

where the parameters $PVW_i \bullet BT_w$ and $PVW_i \bullet ET_w$ are defined as

$$PVW_i \cdot BT_w = \min_{k=1\,to\,m}\left(PVW_{i,k} \cdot ST_w\right) \tag{2.20}$$

$$PVW_i \cdot ET_w = \max_{k=1\,to\,m}\left(PVW_{i,k} \cdot ST_w + PVW_{i,k} \cdot W_w^{0/1}\right) \tag{2.21}$$

2.5 Error Probability Estimation Algorithm

As the major step of SER analysis, the procedure of error probability computation is shown in Algorithm 2.1. Generally, the soft error rate of a given combinational circuit is computed by following the steps below:

1. *Circuit Graph Construction (CG)*: All circuit nodes from PO to every reachable PI are included in a tree while the PO is its root and a subset of PI is its leaves. This tree is extracted using the backward level-ordered algorithm [14] (line 2).

2. *Levelizing*: The circuit nodes in the CG of each PO are levelized using the topo-logical sorting algorithm [14]. Please note that the traditional topological sorting is performed with a modification on the edge direction of the circuit graph; that is, if there is an edge (u, v) in the circuit graph, then u appears after v in the list (line 2).

3. *PVW and Error Probability Computation*: For the nodes which are directly con-nected to POs, the PVW set contains one PVW which is determined by the ini-tialization step using Table 2.1 (line 4). For each circuit node CN_i, the estimation procedure of $P\left(E_i^j\right)$ (and also $P\left(E_i^{j_n}\right)$) is performed considering the number of elements of the PVW set (lines 9–13). After calculating the error probability of the nodes connected to POs, the PVW set of the fan-in gates of the circuit nodes (i.e., the gates which are connected to the gate) are computed using the proposed computation model in line 15. This cycle of the estimation procedure of $P\left(E_i^j\right)$ and the PVW computation is repeated level by level until all gates in all levels are visited. So, the error probability of the combinational circuit is gradually computed based on the error probability of the circuit nodes considering (2.12)–(2.16). SER of the circuit is finally determined by (2.5), (2.2), and (2.1).

Algorithm 2.1 Error Probability Estimation
1: **Input**: Gate-level Netlist, Gate Delays (t_p, t_{rf}, tp_{hl}, tp_{lh}), FF timings (t_s, t_h), T_{clk}
2: Extract the circuit graph (CG) and levelize it
3: **For** each CN of POs input
4:             PVW Initialization // Table 2.1
5: **End For**
7: **For** each level
8:             **For** each CN_i
9:                 Compute E_i for one PVW // Section 2.4
10:                **If** N(PVW_i) > 1 // more than one PVW
11:                    Compute E_i for more than one PVW // Section 2.4.1
12:                **End If**
13:                **For** all fan-in (CN_k) of CN_i
14:                    Compute PVW // Table 2.2
15:                    Add to PVW_k
16:                **End For**
17:             **End For**
18: **End For**

2.6 Experimental Results

In this section, we investigate the efficiency and accuracy of the proposed method comparing with a Monte Carlo-based fault injection technique as the reference model.

2.6.1 Reference Model

A Monte Carlo-based fault injection simulation (MCFI) technique is used as a reference model. In MCFI, SETs with specific initial widths are injected at the output of random fault sites (i.e., circuit gate outputs) at a random time within the clock period. During the fault injection, SETs are propagated to FFs at POs using timing-accurate simulations (i.e., the timing of circuit paths is considered to affect the signals in the circuit while the SETs are injected and propagated to FFs). MCFI terminates if the maximum variance of the estimated value becomes 2% while the confidence level is 99%.

2.6.2 Accuracy/Runtime Investigation

The proposed SER estimation method has been implemented with C++ and applied to the well-known combinational circuits in ISCAS'85 benchmark suite for 45 nm BPTM library technology [15]. All simulations have been run on a Linux machine with a Pentium Core Duo (2.8 GHz) processor and 4 GB RAM. In order to find the error probability of the circuits using the proposed method, the initial width of SETs is set to 200 ps while the clock cycle period (T_{clk}) used is 1000 ps. The setup (t_s) and hold (t_h) times for the flip-flops at POs are set to 10 ps.

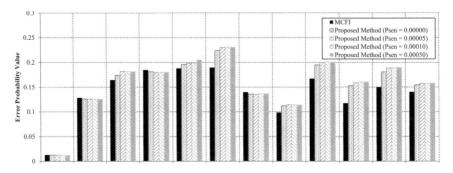

Fig. 2.4 Comparison of the accuracy of the proposed approach with the MCFI technique considering different P_{sen} thresholds (P_{sen}) for PVW screening

In general, the accuracy comparison can be presented in two ways: (1) relative difference of the obtained error probabilities and (2) absolute difference between the calculated error probabilities. Reporting the relative inaccuracy in comparing SER estimation results does not show how the proposed approach follows the reference model [12]. Hence, as shown in Fig. 2.4, we use the absolute difference of the obtained error probabilities to compare the accuracy of the proposed method and MCFI technique. The results are reported in terms of the error probability per gate obtained by each method. The error probability is evaluated considering different P_{sen} threshold (P_{sen}) values used for PVW screening [6] during PVW computations. The value of the absolute average bar (Abs Avg bar) is obtained by computing the average of the absolute difference between the error probability obtained by the proposed method and MCFI technique. As can be seen in Fig. 2.4, the average inaccuracy of the proposed method is about 0.02 compared to MCFI technique. Considering the obtained results of the proposed method and MCFI, the inaccuracy increases for the circuits including larger number of reconvergent paths (e.g., c1908 with more than 11,000 reconvergent paths in comparison with its similar-sized circuit (c1355 and c2670)). Please note that in the view of relative inaccuracy, the error probability of the proposed method has a difference of about 13% on average and the maximum deviation is about 36% which is related to c6288 with more than 35,000 reconvergent paths. Moreover, P_{sen} threshold values chosen for PVW screening do not significantly affect the accuracy of the proposed method; that is, the maximum difference between applying PVW screening with $P_{sen} = 0.00050$ and ignoring this technique ($P_{sen} = 0.0000$) is about 0.01.

Figure 2.5 shows the runtime of MCFI technique and the proposed method considering different cases of PVW screening. Note that the Y-axis is logarithmic in this figure. Compared to MCFI, the proposed method including PVW screening method provides about four orders of magnitude speedup. Using PVW screening, the proposed method becomes five orders faster than MCFI. Moreover, the algorithm acceleration which is achieved by PVW screening technique increases in the circuits with larger number of reconvergent paths. The reason is that reducing the number of PVWs in reconvergent cases exponentially reduces the number of different combinations of PVWs.

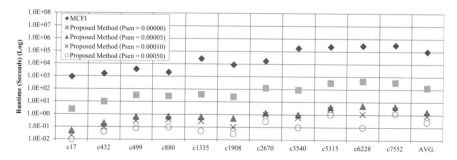

Fig. 2.5 Execution times for the proposed approach and MCFI to compute the design error probability based on different P_{sen} thresholds (P_{sen}) for PVW screening

2.7 Conclusions

In this chapter, we presented a methodology for soft error rate estimation of combinational circuits. Using a vulnerability analysis technique and probabilistic computation model, the effects of all masking factors are considered in the SER estimation of the circuits without any dependency on the initial width of SETs. We have investigated the accuracy and runtime of the proposed approach by comparing it to Monte Carlo-based fault injection simulations.

References

1. Rajaraman R., Kim J. S., Vijaykrishnan N., Xie Y., Irwin M. J.: 'SEAT-LA: A Soft Error Analysis Tool for Combinational Logic', Proc. Int. Conf. VLSI Design (VLSID), 2006, pp. 499–502.
2. Yu-Hsin K., Huan-Kai P., Wen C.H.: 'Accurate Statistical Soft Error Rate (SSER) Analysis Using a Quasi-Monte Carlo Framework with Quality Cell Models', Proc. Int. Symp. on Quality Electronic Design (ISQED), 2010, pp. 831–8.
3. Krishnaswamy S., Plaza S. M., Markov I. L., Hayes J. P.: 'Signature-based SER Analysis and Design of Logic Circuits', IEEE Trans. Computer-Aided Design of Integ. Cir. Sys (TCAD), 2009, 28, (1), pp. 74–86.
4. Zhang M., Shanbhag N.R.: 'Soft-Error-Rate-Analysis (SERA) Methodology', IEEE Trans. Computer-Aided Design (TCAD), 2006, 25, (10), pp. 2140–55.
5. Costenaro E., Alexandrescu D., Belhaddad K., Nicolaidis M.: 'A Practical Approach to Single Event Transient Analysis for Highly Complex Design', Journal of Elect. Test. (JETTA), 2013, 29, (3), pp. 301–315.
6. M. Raji, H. Pedram and B. Ghavami, 'Soft error rate estimation of combinational circuits based on vulnerability analysis,' in IET Computers & Digital Techniques, vol. 9, no. 6, pp. 311–320, 11 2015.
7. Chang A.C.-C., Huang R.H.-M., Wen C.H.-P.: 'CASSER: A Closed-Form Analysis Framework for Statistical Soft Error Rate', IEEE Trans. on Very Large Scale Integ. Sys. (TVLSI), 2013, 21, (10), pp. 1837–1847.
8. Berkeley Predictive Technology Model: http://ptm.asu.edu/
9. Maharrey, J.A., Quinn, R.C., Loveless, T.D., et al.: 'Effect of Device Variants in 32 nm and 45 nm SOI on SET Pulse Distributions', IEEE Trans. on Nuclear Science (TNS), 2013, 60, (6), pp. 2586–2594.
10. Rossi, D., Omaña, M., Metra, C., Paccagnella, A.: 'Impact of Aging Phenomena on Soft Error Susceptibility', Proc. of IEEE Symp. on Defect and Fault Tolerance in VLSI and Nanotechnology Systems (DFT), 2011.
11. Zivanov N.M., Marculescu D.: 'Circuit Reliability Analysis using Symbolic Techniques', IEEE Trans. Computer-Aided Design of Integ. Circ. and Sys. (TCAD), 2006, 25, (12), pp. 2638–2649.
12. Asadi H., Tahoori M.B., Fazeli M., Miremadi S.G.: 'Efficient Algorithms to Accurately Compute Derating Factors of Digital Circuits', Microelec. Rel., 2012, 52, (1) pp. 1215–1226.
13. Mohyuddin N., Pakbaznia E., Pedram M.: 'Probabilistic Error Propagation in Logic Circuits Using the Boolean Difference Calculus', Int. Conf. Computer Design (ICCD), 2008, pp. 7–13.

14. NARASIMHAM, B., BHUVA, B.L., SCHRIMPF, R.D. ET AL.: 'CHARACTERIZATION OF DIGITAL SINGLE EVENT TRANSIENT PULSE-WIDTHS IN 130-NM AND 90-NM CMOS TECHNOLOGIES', IEEE TRANS. ON NUCLEAR SCIENCE (TNS), 2007, 54, (6), PP. 2506–2511.
15. PAPOULIS A., PILLAI S.U., PROBABILITY, 'RANDOM VARIABLES, AND STOCHASTIC PROCESSES', (TATA MCGRAW-HILL EDUCATION PRESS, 2002).

Chapter 3
Process Variation-Aware Soft Error Rate Estimation Method for Integrated Circuits

3.1 Introduction

As technology scales further, variations in cell geometries become prominent as well. According to [1], the reliability of a particle strike transient fault for a cell is sensitive to its geometry size, especially to transients of small charges. Therefore, both process variation impact and soft error effect need to be integrated for accurately estimating SER of nanoscale circuit designs.

In this chapter, we present a paradigm shift for statistical soft error rate estimation of combinational circuits [2]. To this end, we introduce a concept called statistical vulnerability window (SVW) which is a unified variation-aware inference of various constraints necessary for a SET to cause soft errors in the circuit considering all three masking factors. SVW of the circuit nodes (i.e., gate outputs) explicitly indicate the statistical timing and electrical constraints that a SET needs to satisfy in order to be latched in a memory element as a soft error. Besides the timing and electrical masking factors, we take advantage of Boolean difference calculus in order to compute the logical masking factor and integrate its computation inside the SVW parameter computation process. Beginning from a primary output (PO) and using a level-order backward traversal, SVWs for each circuit node are calculated using a probabilistic computation model. The statistical SER of the circuit is gradually estimated considering a mathematical formulation of SER based on SVW parameters.

The rest of this chapter is organized as follows. Section 3.2 gives a general overview of the introduced SER estimation method. In Sect. 3.3, we describe the details of SVWs. Section 3.4 presents in more detail the mathematical model that lies behind the SER framework. Section 3.5 presents the SER estimation algorithm and Sect. 3.6 reports the simulation results. Finally, this chapter is concluded in Sect. 3.7.

© Springer Nature Switzerland AG 2021
B. Ghavami, M. Raji, *Soft Error Reliability of VLSI Circuits*,
https://doi.org/10.1007/978-3-030-51610-9_3

3.2   Statistical Soft Error Rate Estimation Framework

Figure 3.1 shows the overall process of the introduced method which comprises three main components: (1) pre-characterization process, (2) level-by-level backward netlist traversing, and (3) SER computation.

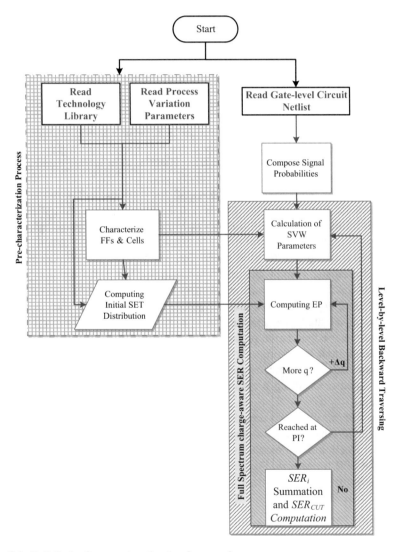

Fig. 3.1   Statistical soft error rate estimation framework

3.2.1   Pre-characterization Process

In order to obtain the parameters necessary for modeling the initial SET pulse width, we conducted the following experiments. For each gate type, Monte Carlo HSPICE simulations were run on a chain of gates of that type. A current source was connected to the output of the first gate in the chain to generate SET. In order to determine the distribution of initial SET pulse width originating at each circuit gate output, parameters which determine the electrical and physical properties between gates and SET width such as output loads, charge strength, and type of driving gate are changed in a sufficient number of runs. Maximum likelihood estimation methodology [3] is used to find the parameters of SET pulse width distributions.

In order to find the timing characteristics of FF, we take advantage of codependent setup and hold time characterization method (CSHT) proposed in [4]. In this method, probability mass function (PMF) of CSHT contours is determined by considering the probability density functions of sources of variability in the first phase. In the second step, PMF of the required setup and hold times for each flip-flop in the design are computed based on a numerical backward Euler-based search.

3.2.2   Level-by-Level Backward Netlist Traversing

Given a multilevel combinational circuit composed of logic gates, we start from each PO and move backwards to the primary inputs (PIs) using a level-order traversal (Fig. 3.2). We calculate the SVWs of circuit node using the statistical precharacteristic data. The error probability of each gate is evaluated using the computed parameters of SVWs and the SET distribution parameters. The SER contribution of each gate circuit (SER_i) is evaluated based on the computed error probability. The process of SER calculation is continued until all gates connected to all POs are visited.

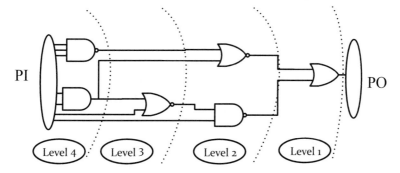

Fig. 3.2 Level-by-level backward netlist traversing

3.2.3   Full-Spectrum Charge Collection-Aware SER Computation

Given the circuit under test (CUT), the overall SER (SER_{CUT}) is estimated as [5, 6]

$$SER_{CUT} = \sum_{i=1}^{n_{Node}} SER_i \tag{3.1}$$

where n_{Node} is the total number of possible struck nodes (assuming every gate output) in the CUT and SER_i is the SER estimated for each circuit node.

Each SER_i can be further elaborated into

$$SER_i = \int_{q=0}^{q_{max}} \left(R(q) \times EP(i,q) \right) dq \tag{3.2}$$

where $EP(i,q)$ shows the probability of a SET originating from a collection charge with strength q at struck node i being latched by at least one flip-flop (FF).

Note that q and q_{max} denote the variables for the collection charge on one cell and the maximum charge value from the environment, respectively. $R(q)$ denotes the effective frequency of a particle strike of charge q in unit time according to [1] as $R(q) = F \times K \times A \times 1/Q_s \times \exp(-q/Q_s)$, where F, K, A, and Q_s are neutron flux rate, a technology-independent fitting constant, the circuit area, and the charge collection slope, respectively.

The step of computing each node SER is composed of two points. First, for a practical SER analysis, the continuous integration in (3.3) is often approximated by a sum of discretized charges [7], that is,

$$SER_i = \sum_{k=1}^{n} R(q_k) \times EP(i,q_k) \tag{3.3}$$

where $q_k = k \times (q_{max} - q_{min})/n$. It is notable that our work is focused in describing the SER estimation method and does not consider the SET generation process. In fact, finding of q_{max} and q_{min} for the charge collection is performed in pre-characterization step. The affective factor in the estimation runtime is the number of discretizing levels (i.e., n). Hence, choosing a suitable value for n requires a trade-off between the accuracy and runtime. The runtime overhead in this trade-off is more considerable for traditional SER estimation methods since they have to repeat all steps of computation procedures for each q (i.e., n in (3.3)). However, the introduced method saves the runtime by shortening this loop.

The second point about SER computation step is about the error probability calculation. $EP(i,q)$ can be calculated as

$$\text{EP}(i,q) = 1 - \left(\prod_{j=0}^{n_{FF}} \left(1 - \text{EP}(i,j,q)\right) \right) \tag{3.4}$$

where n_{FF} denotes the total number of FF in the CUT and $\text{EP}(i,j,q)$ denotes the probability of the latching an error in FF j due to a SET in CN_i with particle strength of q. In order to reduce the complexity of the estimation process, we assume that different POs of the circuit are independent [8, 9]. $\text{EP}(i,j,q)$ can be further computed as

$$\text{EP}(i,j,q) = P_1(i,j) \times P_{e-t}(i,j,q) \tag{3.5}$$

where $P_1(i,j)$ represents the probability that SET is propagated through the logical path from the circuit node CN_i to PO j and $P_{e-t}(i,j,q)$ shows the probability of meeting the electrical and timing constraints by a SET (induced by a charge q at CN_i) to be latched as a soft error in PO j. Calculation of P_1 and P_{e-t} is performed using the calculated parameters of SVW in each gate which will be described in the next section.

3.3   Statistical Vulnerability Window (SVW)

Considering three masking factors (logical, electrical, and timing), we introduce SVW which is an inference of the requirements that should be met by a SET in a given circuit node in order to be latched as a soft error in the presence of process variation. In the following, we firstly introduce the details of SVW and then we describe how to calculate SVW parameters for different logical gates in the circuit.

3.3.1   SVW Parameters

Conceptually, SVW is a metric which shows the circuit vulnerability to the SETs which can occur in a given circuit node (CN_i) and may finally result in soft errors in a memory element j. A SVW can be considered as four-ordered parameters which have been characterized as follows:

- *Output ID* (OID_w): the memory element in which the soft errors will be latched due to the SETs occurred in a given circuit gate output.
- *Starting Time* (ST_w): a random variable which shows the starting point of the time interval of SVW. This parameter shows the constraint on the starting time from which the assumed SET should begin in the given gate in order to be latched as a soft error.

- *Ending Time* (ET$_w$): a random variable which shows the ending point of the time interval of SVW. This parameter shows the constraint on the ending time until which the assumed SET should continue in the given gate in order to be latched as a soft error.
- *Sensitization Probability* (P$_{sen}$): a probabilistic value which is associated with SVW. This value shows the probability of the event that a SET originated in the given gate output finds a logical path to the corresponding output.

Parameters of SVW show the triple constraints which are necessary to be satisfied by a SET to be latched as a soft error. P_{sen} corresponds to the logical masking factor as it shows the probability that SET is not masked due to the logical values of the other inputs of the gate propagating it through the logical path from the circuit gate output to PO. Regarding the timing masking factor, ST$_w$ and ET$_w$ are the representative of the lower and upper limits of the statistical time interval during which the SET should be present such that it will be latched as a soft error. These parameters represent the electrical masking factor as they can be used to extract minimum value for the width of the given SET to result in a soft error in each circuit node. Considering the impact of the process variation, ST$_w$ and ET$_w$ are modeled as random variables as they are related to the timing characteristics of circuit components. Consequently, SVW considers the vulnerability of the circuit nodes to SET and the resulting soft errors.

3.3.2 SVW of POs

In order to calculate the SVWs for all circuit nodes, we firstly initiate the parameters of the SVW for each PO and then traverse the circuit graph backward and compute SVW parameters for all other circuit nodes.

Without loss of generality, we assume that there is a flip-flop in PO j with setup time (t_s) and hold time (t_h) and the clock frequency is equal to T_{clk}. Since we are interested in the occurrence of a soft error due to propagation of a SET in the time interval between two rising (falling) edges of the clock signal, we can take [0, T_{clk}] as the interval of observation. For a signal to be latched, it needs to be stable from the setup time (t_s) before the rising edge of the clock and remain stable for an interval until the hold time (t_h) after the rising edge of the clock. However, the process variations cause timing characteristics of FFs (i.e., t_s and t_h) to vary and hence we propose to take a statistical approach to model them. We use pre-characterizations to find the distribution of FF timing quantities, i.e., $t_s \sim \mathcal{N}\left(\mu_{ts}, \sigma_{ts}^2\right)$ in which μ_{ts} and σ_{ts}^2 are the setup time random variable mean and standard deviation and $t_h \sim \mathcal{N}\left(\mu_{th}, \sigma_{th}^2\right)$ where μ_{th} and σ_{th}^2 are the mean and standard deviation of hold time random variable.

Considering this idea, the initiation value assigned to the parameters of SVW corresponding to PO j and also a justification for our value assignments are as follows:

- $OID_w = j$: the circuit node is connected to the PO j.
- $ST_w \sim \mathcal{N}\left(\mu_{STw}, \sigma^2_{STw}\right)$ where $\mu_{STw} = T_{clk} - \mu_{ts}$ and $\sigma^2_{STw} = \sigma^2_{ts}$: the SET originated at the input of FF in PO j should start (reaches the switching threshold of the FF in PO j) before the time $T_{clk} - t_s$; otherwise it will not be considered as a valid value by FF in PO j and will be masked through timing masking factor [1].
- $ET_w \sim \mathcal{N}\left(\mu_{ETw}, \sigma^2_{ETw}\right)$ where $\mu_{ETw} = T_{clk} + \mu_{th}$ and $\sigma^2_{ETw} = \sigma^2_{th}$: the SET originated at the input of FF in PO j should continue onto $T_{clk} + t_h$ which means that its width should be more than $t_s + t_h$; otherwise it will be masked through timing masking factor [1].
- $P_{sen} = 1$: there is no gate in the path of occurred SET to the PO; there is certainly a path to PO.

A SET can be either a positive pulse (i.e., logical value changes are 0-1-0) or a negative one (i.e., a series of 1-0-1 change in logical values). However, the type of SET does not impose considerable difference in the corresponding computations. Hence, we do not limit the computations to any specific SET type and meanwhile, wherever it is necessary, we provide additional descriptions.

3.3.3 SVW of Internal Circuit Nodes

After initializing SVW parameters at each POs, the parameters of SVWs for other circuit nodes are calculated level by level backward to PIs (Fig. 3.3), as follows:

- OID_w parameters remain unchanged.
- ST_w and ET_w parameters (μ and σ components) are computed using *Statistical SUBtraction* (SSUB) operation [3] resulting in a normally distributed random variable. The parameters of this random variable are computed as

$$SSUB\left(ST_{i1}, D_{Gi1}\right) \sim \mathcal{N}\left(\mu_{ST_{i1}-D_{Gi1}}, \sigma^2_{ST_{i1}-D_{Gi1}}\right)$$

$$\mu_{ST_{i1}-D_{Gi1}} = \mu_{ST_{i1}} - \mu_{D_{Gi1}}$$

$$\sigma^2_{ST_{i1}-D_{Gi1}} = \sigma^2_{ST_{i1}} + \sigma^2_{D_{Gi1}}$$

$$(3.6)$$

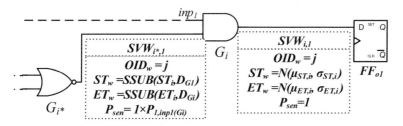

Fig. 3.3 SVW parameter initiation and computation

- Note that ST_{i1} and D_{Gi1} are independent and uncorrelated random variables ($\rho_{ST_{i1},D_{Gi1}} = 0$).
- For a positive SET $\left(\mu_{D_{Gi1}}, \sigma^2_{D_{Gi1}}\right)$ shows the distribution of rising delay of gate G_{i1} and for a negative SET, it represents falling delay distribution.
- SSUB(ET_{i1}, D_{Gi1}) can be computed similarly:

$$\text{SSUB}\left(ET_{i1}, D_{Gi1}\right) \sim \mathcal{N}\left(\mu_{ET_{i1}-D_{Gi1}}, \sigma^2_{ET_{i1}-D_{Gi1}}\right)$$

$$\mu_{ET_{i1}-D_{Gi1}} = \mu_{ET_{i1}} - \mu_{D_{Gi1}}$$

$$\sigma^2_{ET_{i1}-D_{Gi1}} = \sigma^2_{ST_{i1}} + \sigma^2_{D_{Gi1}} \tag{3.7}$$

- Note that ET_{i1} and D_{Gi1} are independent and uncorrelated random variables ($\rho_{ET_{i1},D_{Gi1}} = 0$).
- For a positive SET $\left(\mu_{D_{Gi1}}, \sigma^2_{D_{Gi1}}\right)$ shows the distribution of falling delay of gate G_{i1} and for a negative SET, it represents the rising delay distribution.

- Computation of P_{sen} is similar to the gate error model used in [9] using Boolean difference calculus.

  - It shows the probability of propagating SET through the current gate. In the example shown in Fig. 3.2, it is computed as

$$P_{sen,SVWi2,1} = P_{sen,SVWi1,1} \cdot P_{1,in1Gi2} \tag{3.8}$$

where $P_{sen,\,SVWi2,\,1}$ and $P_{sen,\,SVWi2,\,1}$ show P_{sen} parameters of $SVW_{i2,\,1}$ and $SVW_{i1,1}$, respectively. $P_{1,\,in1Gi2}$ represents the signal probability of input in1 of AND gate G_{i1}. For other cases and type of gates, the computation can be handled similarly.

It is notable that we consider the impacts of electrical masking during computing ST_w and ET_w using a worst-case approximation. For a positive SET, we consider the rising delay of gate G_{i1} for ST_w and the falling delay for ET_w. This will predict the most electrical attenuation which will affect the SET propagating through a given gate. The experimental results show that it does not considerably affect the accuracy of the SER estimation method.

3.4   Error Probability Computation

Considering (3.6), it is required to evaluate P_1 and P_{e-t} to find the error probability of the circuit node. $P_1(i, j)$ represents the probability that SET is propagated through the logical path from the circuit node CN_i to PO j. Hence, it is equal to parameter P_{sen} which is computed in each SVW. $P_{e-t}(i, j, q)$ shows the probability of meeting the electrical and timing constraints by a SET (induced by a charge q at CN_i) to be latched as a soft error (in PO j). This can be decomposed into

$$P_{e-t}(i,j,q) = P_{\text{latching}}(i,j,\text{IS}(i,q))$$
$$= P_{\text{latching}}(i,j,W_{\text{SET}_i}) \tag{3.9}$$

where $P_{\text{latching}}(\bullet)$ denotes the timing masking effects (i.e., latching window constraint), $\text{IS}(i,q)$ shows the computation model used for initial width analysis at node CN_i induced by charge q, and W_{SET_i} represents the width of SET originated at CN_i. Please note that we use a pre-characterization procedure to find the distribution of W_{SET_i} through $\text{IS}(i,q)$.

In the following, we show how to find the P_{latching} using the parameters of SVW computed for CN_i. We firstly present the deterministic mathematical model of finding latching probability and then extend it into the statistical approach. Reconvergent paths are also addressed at the end of this section.

3.4.1 Mathematical Model

In the presented approach, we have determined the necessary constraint which should be satisfied for appearing soft errors by a SET in CN_i. Here, we show how these constraints are determined by the SVW parameters in each circuit node.

In order to be latched at the PO j, the SET occurring in the CN_i should satisfy the setup time and hold time conditions as it has been assumed that there is a FF in PO j. Considering the definition of SVW, this condition implies the necessary conditions for the parameters of the SET at CN_i as follows:

- The starting time of SET should be less than the parameter ST_w of the computed SVW.
- The ending time of the SET at CN_i has to be larger than parameter ET_w.

So, for the starting time (t_{s,SET_i}) and ending time of the SET with width W_{SET_i} in CN_i, we have

$$t_{s,\text{SET}_i} < ST_w \tag{3.10}$$

$$t_{s,\text{SET}_i} + W_{\text{SET}_i} > ET_w \tag{3.11}$$

Thus, we can express the condition by

$$t_{s,\text{SET}_i} \in \left[ET_w - W_{\text{SET}_i}, ST_w \right] \tag{3.12}$$

In addition, for the width of the SET, we have

$$ET_w - ST_w < W_{\text{SET}_i} < \tau \tag{3.13}$$

where τ is the maximum possible width for SETs in a specific technology node [10, 11].

SO, P_{latching} can be expressed as

$$
\begin{aligned}
P_{\text{latching}}\left(i,j,q\right) &= P\left(t_{s,\text{SET}_i} \in \left[\text{ET}_w - W_{\text{SET}_i},\text{ST}_w\right] \cap W_{\text{SET}_i} > W_w\right) \\
&= P\left(t_{s,\text{SET}_i} \in \left[\text{ET}_w - W_{\text{SET}_i},\text{ST}_w\right] \cap \left(\bigcup_k W_{\text{SET}_i} = W_{\text{SET}_i,k}\right)\right) \\
&= \sum_k \left(P\left(t_{s,\text{SET}_i} \in \left[\text{ET}_w - W_{\text{SET}_i},\text{ST}_w\right] | W_{\text{SET}_i} = W_{\text{SET}_i,k}\right) \cdot P\left(W_{\text{SET}_i} = W_{\text{SET}_i,k}\right)\right)
\end{aligned}
\tag{3.14}
$$

where $\{W_{\text{SET}_i,k}\}$ is the set of possible SET widths which may occur in CN_i. Please note that $P\left(W_{\text{SET}_i} = W_{\text{SET}_i,k}\right)$ is determined by $IS(i,q)$ in each circuit node.

Similar to [1, 8], we assume that t_{s,SET_i} is uniformly distributed in the interval $(T_1, T_1 + T_{\text{clk}})$ (T_1 equals the maximum delay from PIs to CN_i). Thus, in the worst case when, for a given SET duration $W_{\text{SET},k}$, the interval $\left[\text{ET}_w - W_{\text{SET}_i},\text{ST}_w\right]$ lies inside it, the probability of the first term in sigma is

$$
\begin{aligned}
& P\left(t_{s,\text{SET}_i} \in \left[\text{ET}_w - W_{\text{SET}_i},\text{ST}_w\right] | W_{\text{SET}_i} = W_{\text{SET}_i,k}\right) \\
& = \frac{W_{\text{SET}_i,k} - \left(\text{ET}_w - \text{ST}_w\right)}{T_{\text{clk}}}
\end{aligned}
\tag{3.15}
$$

assuming $\text{ET}_w - \text{ST}_w < W_{\text{SET}_i} < \tau$.

Equations (3.14) and (3.15) show how to find the probability of latching a SET which is originating in CN_i at PO j.

In Eq. (3.15), $W_{\text{SET}_i,k}$, ET_w, and ST_w are normally distributed random variables as described before. The value of $\text{ET}_w - \text{ST}_w$ can be calculated using a SSUB on ST_w and ET_w random variables resulting into another normally distributed random variable indicated by $\mathcal{W}_w$. In order to evaluate the timing masking effect, we define a new random variable as $w = W_{\text{SET}_i,k} - \mathcal{W}_w$ to compute Eq. (3.15) [7] as

$$
P\left(t_{s,\text{SET}_i} \in \left[\text{ET}_w - W_{\text{SET}_i},\text{ST}_w\right] | W_{\text{SET}_i} = W_{\text{SET}_i,k}\right) = \frac{1}{T_{\text{clk}}} \int_0^{\mu_w + 3\sigma_w} w \times P\left(w > 0\right) dw \tag{3.16}
$$

where $\propto_w$ and σ_w^2 are the mean and standard deviation of w, respectively. w is computed by SSUB operation on $W_{\text{SET}_i,k}$ and $\mathcal{W}_w$ random variables.

It is notable that $w > 0$ in Eq. (3.16) shows the lower bound of initial pulse width (which determines the lower bound of charge q) that is necessary to be considered in each circuit node. The upper bound of pulse widths is set to τ as the maximum possible width for SETs in a specific technology node.

3.4.2 Reconvergent Paths

The equations in Sect. 3.4.1 calculate $P_l(i, j_n)$ and $P_{e-l}(i, j_n, q)$ for the case with one SVW in CN_i regarding PO j. However, there may be more than one SVW computed in CN_i. This situation occurs when there is more than one path from PO j to CN_i; that is, there is a reconvergent path. Figure 3.4 shows an example of a circuit with a reconvergent path. We call a circuit node with a reconvergent fan-out as the *source node* and the circuit node in which the logical paths will reach together is called as the *sink node*.

Considering Fig. 3.4, the SVWs of the corresponding circuit node of G_5 are calculated based on the SVWs of the circuit node related to G_2, G_3, and G_4. Hence, there are at least two SVWs for CN related to G_5. So, the number of SVWs is more than one in the source nodes of reconvergent paths. We gather all SVWs calculated for a given source node CN_i with the same OID j in a set called SVW_i^j, i.e.,

$$SVW_i^j = \left\{ SVW_{i,k}^j |, k = |1, 2|, \ldots |, N_{RP} \right\} \tag{3.17}$$

where N_{RP} is the number of reconvergent paths from source node to the sink node.

In order to address the reconvergent paths case, we propose an approach for evaluating $EP(i, q)$ in CN_i as a source node. Considering Eq. (3.6), it is required to find the probability of latching an error in PO j due to a SET with particle strength of q in CN_i. This value is computed based on the SVWs in SVW_i^j as described in the following.

As the first step, suppose that we have SVW_i^j with two elements in source node CN_i ($SVW_{i,1}^j$ and $SVW_{i,2}^j$). In order to make the analysis more tractable, we define the joint starting and joint ending time of SVWs (JST_w and JET_w, respectively) as

$$JST_w = S\min\left(SVW_{i,1}^j \cdot ST_w, SVW_{i,2}^j \cdot ST_w\right) \tag{3.18}$$

$$JET_w = S\max\left(SVW_{i,1}^j \cdot ET_w, SVW_{i,2}^j \cdot ET_w\right) \tag{3.19}$$

where SMin and SMax are statistical minimum and maximum operations, respectively [12].

In order to address the reconvergent paths case, we evaluate the probability of $EP(i, j, q)$ (and $P_l(i, j)$ and $P_{e-l}(i, j, q)$) using the similar approach as Sect. 3.4.1.

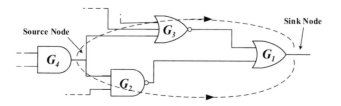

Fig. 3.4 An example of a reconvergent path in a circuit

Regarding evaluating EP(i,j,q), we have two events which may result in a soft error in the circuit, i.e., an event leading to an error by propagating SET through path 1 (with the probability of EP(i_1,j,q)) and an error by propagating SET through path 2 (with the probability of EP(i_2,j,q)). So, using the addition law of probability [3], the probability of EP(i,j,q) can be rewritten as

$$\begin{aligned}
\mathrm{EP}\left(i,j,q\right) &= \mathrm{EP}\left(i_1 \cup i_2,j,q\right) \\
&= \mathrm{EP}\left(i_1,j,q\right) + \mathrm{EP}\left(i_2,j,q\right) - \mathrm{EP}\left(i_1 \cap i_2,j,q\right)
\end{aligned} \tag{3.20}$$

where EP($i_1 \cup i_2,j,q$) denotes the probability of latching an error due to propagating an SET from either path 1 or 2 and EP($i_1 \cap i_2,j,q$) shows the probability of latching an error due to propagating a SET from both paths 1 and 2.

In Sect. 3.4.1, we showed how to compute EP(i_1,j,q) and EP(i_2,j,q). We use the same procedure for evaluating EP($i_1 \cap i_2,j,q$) as Sect. 3.4.1, in addition to the necessary consideration of this case.

Similar to the case of a single SVW, it is required to find logical and electrical masking probabilities, i.e.,

$$\mathrm{EP}\left(i_1 \cap i_2,j,q\right) = P_l\left(i_1 \cap i_2,j\right) \cdot P_{\mathrm{e-t}}\left(i_1 \cap i_2,j,q\right) \tag{3.21}$$

where $P_l(i_1 \cap i_2,j)$ denotes the probability that SET is propagated through both logical paths of 1 and 2 from the circuit node CN$_i$ to PO j and $P_{\mathrm{e-t}}(i_1 \cap i_2,j,q)$ shows the probability of meeting the electrical and timing constraints by a SET (induced by a charge q at CN$_i$) after propagation through both paths 1 and 2 in order to be latched as a soft error (in PO j). In order to compute $P_l(i_1 \cap i_2,j)$ and $P_{\mathrm{e-t}}(i_1 \cap i_2,j,q)$, we take a worst-case approach as follows.

We can decompose $P_{\mathrm{e-t}}(i_1 \cap i_2,j,q)$ into

$$\begin{aligned}
P_{\mathrm{e-t}}\left(i_1 \cap i_2,j,q\right) &= P_{\mathrm{latching}}\left(i_1 \cap i_2,j,\mathrm{IS}\left(i,q\right)\right) \\
&= P_{\mathrm{latching}}\left(i_1 \cap i_2,j,W_{\mathrm{SET}_i}\right)
\end{aligned} \tag{3.22}$$

where P_{latching} denotes the timing masking effects considering the propagation of the SET through both paths. Note that IS(i,q) is similar to the previous case as reconvergent paths do not affect the generation process of a SET. Hence, it is required to handle the evaluation of $P_{\mathrm{latching}}\left(i_1 \cap i_2,j,W_{\mathrm{SET}_i}\right)$ in this case.

Using the similar mathematical model as Sect. 3.4.1, we readdress the corresponding (electrical and timing) constraints for evaluating $P_{\mathrm{latching}}\left(i_1 \cap i_2,j,W_{\mathrm{SET}_i}\right)$ as

$$\mathrm{SVW}_i \cdot \mathrm{JET}_w < W_{\mathrm{SET}_i} < \tau \tag{3.23}$$

$$t_{\mathrm{s,SET}_i} \in \left[\mathrm{SVW}_i \cdot \mathrm{ET}_w - W_{\mathrm{SET}_i}, \mathrm{SVW}_i \cdot \mathrm{JST}_w\right] \tag{3.24}$$

Therefore, the probability of $P_{\mathrm{latching}}\left(i_1 \cap i_2,j,W_{\mathrm{SET}_i}\right)$ can be written as

$$P_{\text{latching}}\left(i_1 \cap i_2, j, W_{\text{SET}_i}\right)$$

$$= P\left(t_{\text{s,SET}_i} \in \left[\text{JET}_w - W_{\text{SET}_i}, \text{JST}_w\right] \cap W_{\text{SET}_i} > \text{PVW}_i \cdot \text{JET}_w\right)$$

$$= P\left(t_{\text{s,SET}_i} \in \left[\text{JET}_w - W_{\text{SET}_i}, \text{JST}_w\right] \cap \left(\bigcup_k W_{\text{SET}_i} = W_{\text{SET}_i,k}\right)\right) \qquad (3.25)$$

$$= \sum_k \left(P\left(t_{\text{s,SET}_i} \in \left[\text{JET}_w - W_{\text{SET}_i}, \text{JST}_w\right] \mid W_{\text{SET}_i} = W_{\text{SET}_i,k}\right) \cdot P\left(W_{\text{SET}_i} = W_{\text{SET}_i,k}\right)\right)$$

where $\{W_{\text{SET}_i,k}\}$ is the set of possible SET widths which may occur in CN_i. The probability of the first term in sigma is

$$P\left(t_{\text{s,SET}_i} \in \left[\text{JET}_w - W_{\text{SET}_i}, \text{JST}_w\right] \mid W_{\text{SET}_i} = W_{\text{SET}_i,k}\right)$$

$$= \frac{W_{\text{SET}_i,k} - \left(\text{JET}_w - \text{JST}_w\right)}{T_{\text{clk}}} \qquad (3.26)$$

It is notable that, considering (3.26), for SETs with the initial width less than $\text{JET}_w - \text{JST}_w$ (i.e., lower bound of pulse width), the reconvergent path case is not required to be considered. Moreover, considering the definition of τ (i.e., upper bound of pulse width), SETs with the initial width larger than τ are assumed to be impossible to occur, i.e., in (3.25), $P\left(W_{\text{SET}_i} = W_{\text{SET}_i,k}\right) = 0$ for $W_{\text{SET}_i,k} > \tau$. Moreover, the computations for reconvergent path case are not required to be considered for the SVWs in which $(\text{JET}_w - \text{JST}_w) > \tau$.

We use an approximation for considering the constraint imposed by logical masking factor. The sensitization probability at source nodes of reconvergent paths should be updated to reflect the phenomenon of reconvergence. For joint SVWs, the result of sensitization probability (i.e., $P_1(i_1 \cap i_2, j)$) is the intersection of the sensitization probabilities of input SVWs, because this condition is equivalent to the SETs passing through both the reconvergent paths from the source to the sink node. So we estimate this value by multiplying P_{sen} of $\text{SVW}_{i,1}^j$ and $\text{SVW}_{i,2}^j$. However, the experimental result in Sect. 3.6 shows that this approximation does not impose much error to the soft error rate estimation method. So (3.22) can be rewritten as

$$\text{EP}\left(i_1 \cap i_2, j, q\right) = P_1\left(i_1 \cap i_2, j\right) \times P_{\text{latching}}\left(i_1 \cap i_2, j, W_{\text{SET}_i}\right)$$

$$= \left(\text{SVW}_{i,1} \cdot P_{\text{sen}} \cdot \text{SVW}_{i,2} \cdot P_{\text{sen}}\right)$$

$$\times \sum_k \left(\frac{W_{\text{SET}_i,k} - \left(\text{JET}_w - \text{JST}_w\right)}{T_{\text{clk}} - W_{\text{SET}_i,k}} \cdot P\left(W_{\text{SET}_i} = W_{\text{SET}_i,k}\right)\right), \qquad (3.27)$$

$$\text{for } \left(\text{JET}_w - \text{JST}_w\right) < W_{\text{SET}_i,k} < \tau$$

By calculating the value of $EP(i_1 \cap i_2, j, q)$ using (3.27) and the value of $EP(i_1, j, q)$ and $EP(i_2, j, q)$ using Eqs. (3.16) and (3.17), $EP(i, j, q)$ can be evaluated as the main part of SER estimation.

The mentioned approach is extensible when there are more than two SVWs. However, there are some considerations which will be mentioned in the next subsection.

3.4.3   Addressing Reconvergent Path Case

In the general case, we assume that there are N_{RP} SVWs in set SVW_i^j; that is, there are N_{RP} reconvergent paths from CN_i as the source node. The error probability of CN_i related to PO j can be computed as

$$EP(i,j,q) = \sum_{k=1}^{N_{RP}} EP(i_k, j, q)$$

$$+ \sum_{m=2}^{N_{RF}} (-1)^{m+1} \sum_{k_1=1}^{N_{RF}} \sum_{k_2=k_1+1}^{N_{RF}} \cdots \sum_{k_m=k_{m-1}+1}^{N_{RF}} EP\left(\bigcap_{l=1}^{m} i_{k_l}, j, q\right) \tag{3.28}$$

where $EP(i_k, j, q)$ is the error probability of CN_i computed considering individual SVWs. Using (3.28), the error probability of CN_i for the nodes with more than two SVWs can be calculated in two steps: (1) the first summation is computed using the parameters of individual SVW, and (2) we use the similar procedure for the general term with m SVWs as the case of a circuit node with two SVWs with some additional considerations as follows:

- Parameters $SVW_i \bullet JST_w$ and $SVW_i \bullet JET_w$ are defined as

$$JST_w = \min\left\{SVW_{i,k} \cdot ST_w \text{ for } k = 1 \text{ to } m\right\} \tag{3.29}$$

$$JET_w = \max\left\{SVW_{i,k} \cdot ET_w \text{ for } k = 1 \text{ to } m\right\} \tag{3.30}$$

- For the general case with m SVWs, considering (3.27), the argument of the second summation can be calculated as

$$EP\left(\bigcap_{l=1}^{m} i_{k_l}, j, q\right) = \prod_l SVW_{i,k_l} \cdot P_{sen}$$

$$\times \sum_k \left(\frac{W_{SET_i,k} - (JET_w - JST_w)}{T_{clk}} \cdot P\left(W_{SET_i} = W_{SET_i,k}\right)\right),$$

$$\text{for } (JET_w - JST_w) < W_{SET_i,k} < \tau \tag{3.31}$$

- The computations for reconvergent path case are not required to be considered for the SVWs for the following case: (1) for SETs with the initial width less than $JET_w - JST_w$ and (2) for the SVWs in which $(JET_w - JST_w) > \tau$.

3.5 Error Probability Estimation Algorithm

As the major step of SER analysis, the procedure of error probability computation is shown in Algorithm 3.1. Generally, the soft error rate of a given combinational circuit is computed by following the steps bellow:

1. *Circuit Graph (CG) Construction*: All circuit nodes from PO to every reachable PI are included in a tree while the PO is its root and a subset of PI is its leaves. This tree is extracted using the backward level-ordered algorithm [13] (line 2).
2. *Levelizing*: The circuit nodes in the CG of each PO are levelized using the topological sorting algorithm [13]. Please note that the traditional topological sorting is performed with a modification on the edge direction of the circuit graph; that is, if there is an edge (u, v) in the circuit graph, then u appears after v in the list (line 2).
3. *SVW and Error Probability Computation*: For the nodes which are directly connected to POs, the SVW set contains one SVW which is determined by the initialization step described in Sect. 3.3.2 (line 4). For each circuit node CN_i, the estimation procedure of $P\left(E_i^j\right)$ (and also $P\left(E_i^{j_n}\right)$) is performed considering the number of elements of the SVW set using the mathematical model presented in Sect. 3.3.3 (lines 9–13). After calculating the error probability of the nodes connected to POs, the SVW set of the fan-in gates of the circuit nodes (i.e., the gates which are connected to the gate) are computed using the computation model presented in Sect. 3.3.3 (line 15). This cycle of the estimation procedure of $P\left(E_i^j\right)$ and the SVW computation is repeated level by level until all gates in all levels are visited. So, the error probability of the combinational circuit is gradually computed based on the error probability of the circuit nodes. SER of the circuit is finally determined by (3.5), (3.4), and (3.2).

Algorithm 3.1 Error Probability Estimation
1: **Input**: Gate-level Netlist, Gate Delays (t_p, t_{rf}, tp_{hl}, tp_{lh}), FF timings (t_s, t_h), T_{clk}
2: Extract the circuit graph (CG) and levelize it
3: **For** each CN of POs input
4:      SVW Initialization // Section 3.3.2
5: **End For**
7: **For** each level of circuit
8:      **For** each CN_i

9:          Compute EP_i for one SVW//Section 3.4.1
10:         **If** N(SVW$_i$) > 1 // more than one SVW
11:             Compute EP_i for more than one SVW // Sections 3.4.2
and 3.4.3
12:         **End If**
13:         **For** all fan-in (CN$_k$) of CN$_i$
14:             Compute SVW // Section 3.3.3
15:             Add to SVW$_k$
16:         **End For**
17:     **End For**
18: **End For**

3.6   Experimental Results

The SER estimation method is implemented in C++ and run on a Linux machine with a Pentium Core i5 (2.4 GHz) processor and 16 GB RAM. The neutron flux rate F was set as 56.5 m^{-2} s^{-1} at sea level [14] and the charge collection slope was set to $Q_s = 10.84$ fC [14]. We assume 10% process variation on electrical characteristics of circuit components (i.e., transistor channel width and length) in the benchmark circuits. SER values are reported in failure in time (FIT), which is defined as one failure in 10^9 h.

Monte Carlo SPICE simulations on ISCAS'85 benchmark circuits are conducted in order to extract delay characteristics of each type of gate in a 45 nm Nangate Open Cell Library [15]. We use the procedure proposed in [16] to obtain the pre-characterization delay data for each gate type. The procedure can be summarized in three steps: (1) a subset of the gates in the circuit in different propagation paths are randomly selected; (2) different output loads composed of the selected gates are arbitrarily considered for each gate in the circuit; (3) the characteristics of the gate delays are extracted by performing Monte Carlo SPICE simulation. In addition, the characteristics of SETs induced by radiation particles with various charge strengths are determined in similar simulation procedure. Using the obtained simulation results, the behavior of circuit elements towards SETs (i.e., initial pulse generation models used in function IS(i, q) for circuit node i with particle charge of q) is extracted in addition to their timing characteristics.

The experiments are divided into two parts. In the first part of the experiments, we verify the introduced SER estimation method and then, in the second part, we compare the presented method with other previous approaches considering full-spectrum charge in SER estimation.

3.6.1 Verification of the SER Estimation Method

Extensive Monte Carlo SPICE (MCS) simulations with 10,000 repetitions are used as a reference method for each benchmark circuit in accuracy verification of the SER estimation method .Each repetition is a process of fault injection (with 10,000 rounds) by fixing process variations into a set of randomly sampled values. MCS runtime increases drastically by the number of gates and the circuit inputs. For example, the runtime of this simulation for even small circuits with 62 gates and 12 inputs requires more than a week, limiting us to perform tests on such small circuits in accuracy verification procedure. Hence, the experiments are performed on five benchmark circuits (b1, b2, b3, c17, Adder2bit, C1, and C2). To consider the impacts of process variations, three parameter variations are considered, i.e., channel length, channel width, and threshold voltage. Different levels of variability in these parameters were explored, i.e., 10%, 20%, and 40% variation ratio ($3\sigma/\mu$). The SER estimation method can be easily extended to include other parameters of variations. Table 3.1 reports the information related to the benchmark circuits in the experiments in the first two columns. The rest of the columns present the SSER analysis results obtained from MCS and the proposed method. It also reports the error between the SSER values, i.e., $\dfrac{SSER - MCS}{MCS}$. The obtained results show that increasing variation ratio from 10% to 40% does not decrease the accuracy of the presented method.

3.7 Conclusion

This chapter presented a framework which considers process variation impacts on SER estimation of digital circuits. In this method, the vulnerability of the circuit gates is determined using a concept called SVW representing all triple masking factors. The main advantage of the introduced method is that it achieves an acceptable accuracy and efficiency trade-off by considering all levels of deposited charges in SER estimation with less runtime compared to the recent similar approaches. We have demonstrated the efficiency of the presented method by applying it on a subset of benchmark circuits. Compared to the Monte Carlo HSPICE simulations, the SER estimation method provides considerable speedup (about five orders of magnitude) with less than 5% accuracy.

Table 3.1 SER measurement results on benchmark circuits considering different variation ratios

Circuit	#Gate(#PI, #PO)	MCS (μFIT)			Proposed (μFIT)			Error (%)		
		10%	20%	40%	10%	20%	40%	10%	20%	40%
INV10	10(1, 1)	31.56	53.652	85.84	31.68	53.92	86.44	0.4	0.5	0.7
DEC2:4	6(2, 4)	84.48	135.16	229.78	89.12	142.87	244.26	5.5	5.7	6.3
MUX4:1	7(6, 1)	57.37	94.66	165.65	58.86	97.02	169.46	2.6	2.5	2.3
COM2	13(4, 1)	36.19	57.18	96.06	37.13	58.49	98.656	2.6	2.3	2.7
c17	12(5, 2)	66.73	117.44	194.95	69.33	122.37	202.75	3.9	4.2	4
ADD2	18(5, 3)	79.85	124.56	219.23	83.36	130.42	228.22	4.4	4.7	4.1
C1	33(7, 4)	106.2	174.16	268.21	113.95	187.75	288.60	7.3	7.8	7.6
C2	62(12, 7)	117.2	194.55	342.41	124.34	206.22	364.32	6.1	6.0	6.4
Average								4.1	4.2	4.2

References

1. A. Dixit and A. Wood, "The Impact of New Technology on Soft Error Rates," International Reliability Physics Symposium (IRPS), pp. 5B.4.1–5B.4.7, 2011.
2. M. Raji and B. Ghavami, A Fast Statistical Soft Error Rate Estimation Method For Nano-scale Combinational Circuits, Journal of Electronic Testing (JETTA), Vol. 32, No. 1, pp. 291–305, 2016.
3. A. Papoulis, S.U. Pillai, Probability, Random Variables, and Stochastic Processes, Tata McGraw-Hill Education, 2002.
4. S. Hatami, H. Abrishami, M. Pedram, "Statistical timing analysis of flip-flops considering codependent setup and hold times," ACM Great Lakes symposium on VLSI (GLSVLSI), pp. 101–106, 2008.
5. M. Raji, H. Pedram, B. Ghavami, "A Practical Metric for Soft Error Vulnerability Analysis of Combinational Circuits," Microelectronics Reliability, Vol. 55, No. 2, pp. 448–460, 2015.
6. M. Raji, H. Pedram, B. Ghavami, "Soft Error Rate Estimation of Combinational Circuits Based on Vulnerability Analysis," IET Computers & Digital Techniques, Vol. 9, No. 6, pp. 311–320, 2015.
7. A. C.-C. Chang, R. H.-M. Huang, C. H.-P. Wen, "CASSER: A Closed-Form Analysis Framework for Statistical Soft Error Rate," IEEE Transactions on Very Large Scale Integration Systems (TVLSI), Vol. 21, No. 10, pp. 1837–1848, 2013.
8. N. Miskov-Zivanov, K.-C. Wu, D. Marculescu, "Process Variability-aware Transient Fault Modeling and Analysis," Int'l Conf. Computer Aided Design (ICCAD), pp. 685–690, 2008.
9. N. Mohyuddin, E. Pakbaznia, M. Pedram, "Probabilistic Error Propagation in Logic Circuits using the Boolean Difference Calculus," International Conference on Computer Design (ICCD), pp. 7–13, 2008.
10. J.A. Maharrey, et al. "Effect of Device Variants in 32 nm and 45 nm SOI on SET Pulse Distributions," IEEE Transactions on Nuclear Science (TNS), Vol. 60, No. 6, pp. 2586–2594, December 2013.
11. B. Narasimham, et al., "Characterization of Digital Single Event Transient Pulse-Widths in 130-nm and 90-nm CMOS Technologies," IEEE Transactions on Nuclear Science (TNS), Vol. 54, No. 6, 2007.
12. Saralees Nadarajah and Samuel Kotz, "Exact Distribution of the Max/Min of Two Gaussian Random Variables," IEEE Transactions on Very Large Scale Integration Systems (TVLSI), Vol. 16, No. 2, pp. 210–212, 2008.
13. T.H. Cormen, C.L. Leiserson, R.L. Rivest, C. Stein, Introduction to Algorithms, MIT Press & McGraw-Hill, 2001, 2nd edition.
14. P. E. Dodd and L. W. Massengill, "Basic Mechanisms and Modeling of Single-Event Upset in Digital Microelectronics," IEEE Transaction on Nuclear Science, Vol. 50, No. 3, pp. 583–602, Jun. 2003.
15. Nangate Inc. (2009). Nangate 45 nm Open Library, Sunnyvale, CA [Online]. Available: http://www.nangate.com/
16. H.K. Peng, C. H.-P Wen, and J. Bhadra, "On Soft Error Rate Analysis of Scaled CMOS Designs: A Statistical Perspective," Int'l Conf. Computer Aided Design (ICCAD), pp. 157–163, 2009.

Chapter 4
GPU-Accelerated Soft Error Rate Analysis of Large-Scale Integrated Circuits

Recently, parallel hardware presents an opportunity to solve the electronic design automation (EDA) challenges and opens up new design automation opportunities which provide orders of magnitude faster executions of EDA algorithms. Commercial circuit simulators like Spectre Accelerated Parallel Simulator from Cadence can provide real spice circuit simulations with a full accuracy using many core CPUs or even clusters. In addition, different hardware platforms such as reconfigurable devices, custom-designed ICs, and graphic processing units (GPUs) are viable alternatives to achieve this acceleration. GPUs are designed to operate in a single-instruction multiple data (SIMD) fashion. In recent years, the application of GPUs for general-purpose computations has been actively explored [1]. In general, these graphics acceleration units process the same operations (i.e., instructions) independently on large volumes of data. Tremendous speedups are gained by using GPUs for problems in EDA including electrical simulation and circuit optimization. Specifically, in [2], a GPU-based small delay fault (SDF) simulation framework is proposed. The main goal of the method presented in [2] is to obtain the test vectors required to find the SDFs induced by process variations and noise.

Exploring of GPU-based parallelism for SER estimation is motivated by the observation that the algorithm repeats a few identical computations on a large number of gates and for multiple iterations. Evidently, such repetitive computations on a large number of objects fit very well with SIMD parallelism.

This chapter presents a fast SER estimation methodology for very large combinational circuits. The method exploits GPU platforms to accelerate the SER estimation process. A multidimensional parallelism approach is presented to provide maximum speedup. Several groups of SETs (fault parallelism) are injected simultaneously to multiple fault locations (structural parallelism), and the SETs in each circuit level are processed in parallel (gate parallelism). Therefore, the SER of a large number of circuit gates is computed by a single-circuit graph traversing. The introduced method accurately estimates the circuit SER for a wide range of SETs with different characteristics.

© Springer Nature Switzerland AG 2021
B. Ghavami, M. Raji, *Soft Error Reliability of VLSI Circuits*,
https://doi.org/10.1007/978-3-030-51610-9_4

This chapter is organized as follows: Section 4.1 provides an overview of the proposed approach. SER analysis process is discussed in details in Sect. 4.2. Section 4.3 demonstrates the three dimensions of parallelism exploited here. Section 4.4 describes the introduced SER estimation process in detail. Section 4.5 presents the results of proposed technique and finally the conclusions are made in Sect. 4.6.

4.1   Approach Overviews

Figure 4.1 shows the overall parallel SER estimation approach. This approach consists of sequential and parallel tasks. The sequential tasks (white boxes), which involve cell pre-characterization and scheduling initializing, are performed on the CPU side. The parallel tasks (shaded box), which contain the main simulation loop, are performed on the GPU side.

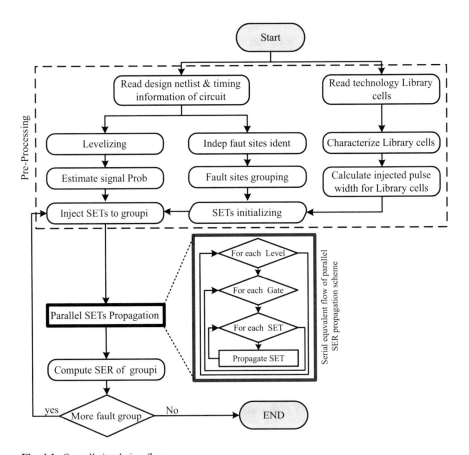

Fig. 4.1 Overall simulation flow

The approach simultaneously employs three dimensions of parallelism, gate parallelism, fault parallelism, and structure parallelism. Gate parallelism is exploited by evaluating all gates within a same level in parallel. Multiple independent SETs are injected into a fault location at the same time to provide fault parallelism. Structure parallelism refers to injecting SETs into a set of structural independent fault sites (gates with no overlapping fan-out cone) at the same simulation time.

The first part of the algorithm (upper part) represents the preprocessing initializing steps of the proposed approach. The inputs of the framework are (1) design netlist and timing information, (2) particle charge energy, and (3) technology library cells. During the pre-characterization step, the pulse width of the injected SETs for different cells is extracted using the particle energy and the technology library cells. After reading a design netlist, the circuit is topologically ordered and the signal probabilities (SP) at the output of all logic gates are computed using 0-algorithm [3]. It should be noted that SP values can be imported as they are usually computed for power analysis purposes. However, if they are not available from previous design steps, SPs can be computed using either the probabilistic or Monte Carlo simulation methods. Also, the fan-out cone of each gate is extracted to identify the independent fan-out cones. Then independent gates are divided into a set of groups in order to inject the faults simultaneously.

The second part of the algorithm contains the main estimation process composed of injecting parallel SETs into parallel fault sites and propagating SETs through the circuit.

4.2 SER Analysis

In this section, we briefly present the soft error rate analysis method [4] which is accelerated using the proposed acceleration framework. In this approach, soft error analysis of a combinational circuits is built upon a probabilistic framework which incorporates the probability theory, circuit simulation, graph theory, and fault simulation. The overall SER of a circuit under test (CUT) can be defined as the summation of soft errors (SE) resulting from particle hits at each individual gate (C_i) in the circuit, as follows:

$$\text{SER}_{\text{CUT}} = \sum_{i=1}^{N_{\text{gate}}} \text{SE}_{C_i} \tag{4.1}$$

where N_{gate} denotes the total number of gates susceptible to radiation particle hits in the circuit. Note that the transient fault caused by a particle hit may propagate and be captured by different memory elements, leading to numerous soft errors.

Each SE(C_i) can be further formulated by integrating the products of the particle hit rate and the error probability over the range of charge strength from q_{min} to q_{max} as

$$SE_{C_i} = \int_{q_{min}}^{q_{max}} R_{PH}(q) \times P_{err}(C_i, q) dq \tag{4.2}$$

where $P_{err}(C_i, q)$ denotes the probability of a transient fault originating from a collection charge with strength q at hit node C_i being latched by one flip-flop. $R_{PH}(q)$, the particle hit rate, is the effective frequency at which particle with strength q hits the circuit in unit time resulting in charge generation [5]. The error probability $P_{err}(C_i, q)$ depends on all three masking effects illustrated in Fig. 4.1, which can be further decomposed into

$$P_{err}(C_i, q) = \sum_{j=1}^{N_{ff}} P_{logc}(C_i, d_j) \times P_{elec}(C_i, q, d_j) \times P_{latch}(C_i, q, d_j) \tag{4.3}$$

where N_{ff} and d_j represent the total number of flip-flops in the circuit and the jth flip-flop, respectively. P_{logc}, P_{elec}, and P_{latch}, respectively, represent the logic masking probability, electrical masking probability, and timing masking probability. $P_{logc}(C_i, d_j)$ represents the overall logic probability of successfully propagating the transient faults through a path from gate to flip-flop (denoted by $C_i \rightarrow d_j$). $P_{logc}(C_i, d_j)$ is computed considering the signal probabilities of output of gate C_i multiplied by all signal probabilities of non-controlling value of all gates on paths towards d_j, and can be expressed as

$$P_{logc}(C_i, d_j) = P_{sig}(C_i^*) \times \prod_{C_k \in C_i \rightarrow d_j} P_{sig}(C_k) \tag{4.4}$$

where $P_{sig}(C_i^*)$ is the probability of logic-0 (logic-1) when a positive (negative) transient fault is generated at C_i and C_k, which is neither C_i nor d_j, is a gate in the logical path from C_i to d_j ($C_i \rightarrow d_j$). $P_{sig}(C_k)$ represents the signal probability for a non-controlling side input that does not impede a transient fault propagating through gate C_k.

The second term $P_{elec}(C_i, q, d_j)$ in (4.3) denotes the overall probability that SETs which are generated at node C_i with deposited charge of q are propagated to output d_j with recognizable electrical amplitude along all paths assuming that all side inputs always have non-controlling values. The third term $P_{latch}(C_i, q, d_j)$ in (4.3) indicates the probability that the SETs are induced by charge q at node C_i, getting latched into the flip-flop at output d_j considering no logical masking or electrical masking effects. Error latching probability P_{latch} for one flip-flop is defined as follows:

$$P_{latch} = \frac{PW - W}{t_{clk}} \tag{4.5}$$

where PW, W, and t_{clk} denote the pulse width of the arrival transient fault, the latching window (t_{setup}+ t_{hold}) of the flip-flop, and the clock period, respectively. The three masking effects are considered during the proposed estimation approach. Without loss of generality, we employed the proposed techniques in [6] and [7] to model the electrical attenuation of transient faults and analysis timing masking effect, respectively.

For an accurate soft error rate analysis, SER_{CUT} is computed through generating all possible voltage pulses at each node and propagating the SET voltage pulses to the primary outputs (POs) through all possible paths, and getting latched into at least one FF. Note that, in traditional SER estimation approaches, each generated SET pulse at each node needs to be propagated through all possible paths towards the POs, making the propagation process very time consuming even for small- and middle-size circuits, while in the proposed parallel approach, all possible SETs are injected to multiple gates at the same time and the propagation process in each level is performed in parallel. This provides a significant speedup in SER_{CUT} computation.

4.3 Parallel SER Analysis

In this section, we present the parallelization approaches, which are proposed for accelerating the SER estimation process. We show how conventional sequential SER estimation algorithm is mapped onto GPU computation to achieve the demanded speedup.

4.3.1 Gate Parallelism

Gate parallelism is defined as the propagation of SETs through G data-independent gates within a single simulation instance. Two gates are mutually data independent when the inputs of any of the gates do not depend on the output values of the others and vice versa. To provide gate parallelism, a simple topological sort algorithm is performed on the combinational circuit graph, where the first level contains all primary inputs and for the rest of the circuit the fan-in gates of all gates in each level are computed in the previous levels. Therefore, the resulting levelized circuit is the sets of pairwise independent gates in each level and thus it is possible to simulate the entire gates of one level at a time. Figure 4.2 shows the topologically sorted version of the circuit in Fig. 4.1. In Fig. 4.1, suppose that the transient fault generated in G0 in level 1 would reach the inputs of G2 and G3 in level 2 through the G0 → G2 and G0 → G3 paths. Since G2 and G3 are located in the same level, they are data independent and therefore the SETs that appeared in their inputs can be evaluated in parallel. In the proposed method, each thread is responsible for propagating a single SET through a gate in each level. It is notable that the amount of parallelism that can

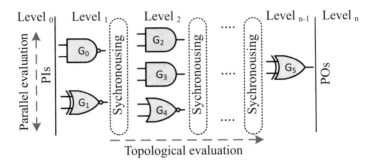

Fig. 4.2 Parallel evaluation of data-independent gates in a topologically ordered circuit

be achieved is limited by the number of gates in each level and the number of gates in each level only depends on the circuit structure.

4.3.2 Parallel Structure

As a main step of all circuit-level SER analysis approaches, after selecting a gate for SET injection, the forward cone (gates which are located between a specific gate and POs) of the selected gate is extracted and the SET propagates through this cone and towards the POs. This process is repeated for all circuit gates, individually.

In order to accelerate SER analysis by providing maximum parallelism and reducing the number of simulation instances, we employ a *multiple independent SET injection (MISI)* scheme. In *MISI*, several SETs are injected into structurally independent fault sites simultaneously. Two fault sites are said to be structurally independent if the generated SETs in these locations will not affect each other when propagating towards POs. Having any interaction, SETs can be injected in these fault sites at the same simulation instance. In Fig. 4.1, G0 and G1 are structurally independent, as the SETs generated in G0 and G1 will not affect each other and propagate to different parts of the circuit which share no common POs. The problem of determining optimal fault sites to simultaneous fault injection in order to provide maximum simulation speedup can be formulated as a *minimum graph coloring problem* [2]. Since finding even the approximate optimal solution for this problem is proved to be NP hard, many heuristic approaches have been proposed for efficiently grouping the fault sites. Here, we apply the greedy-based method presented in [2] to group the independent sites into different sets for large-scale circuits. Site grouping is a pre-processing step which is ran only once on the host system.

4.3.3 Parallel SET

Previous studies have shown that various pulse widths (from less than 100 ps to greater than 700 ps) are generated as a function of particle's LET (linear energy transfer) [8]. It has also been observed that even a particle with a given LET will lead to a broad range of pulse widths [8]. Therefore, it is required to study the effect of a distribution of pulse widths, instead of a single maximum SET pulse width. An accurate SER analysis provides the opportunity to harden the circuit design against soft errors, using much smaller penalties than previously believed.

Considering various SETs with different pulse widths causes the SER analysis complexity to increase quickly, since each gate has to be evaluated for a wide range of SETs. In the proposed method, multiple independent SETs (referred to as *SET group*) with different widths and polarities are injected into each gate at the same time. The proposed parallel SET scheme operates in a data-parallel manner in which each individual thread is responsible for propagating one of the independent SET in a SET group through a gate. Figure 4.3 shows two SET groups which are injected into a site group (G0 and G1) simultaneously and propagated through the circuit. Since all SETs are propagated with only one single pass over the circuit, the proposed method provides high accuracy without any runtime cost. It is notable that the pulse width of the transient fault does not affect the simulation time in the proposed acceleration framework.

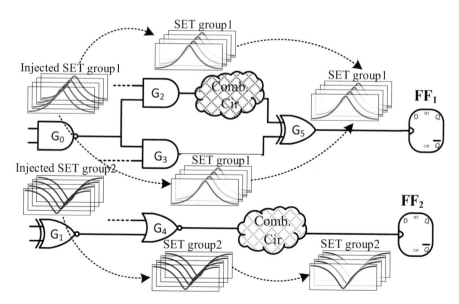

Fig. 4.3 Parallel simulation of multiple SETs

4.4 Simulation

4.4.1 Preprocessing

Simulation begins with the host CPU, which transfers the kernel code and data structures to the GPU and starts the simulation of each cycle. The data is organized as a "C" structure *type struct* which is stored in the global memory of the device for all thread blocks. The circuit design data contains the type and input name for each gate and the circuit gates are organized by their topological order. Transient fault data contains logical probability, start time, pulse width, and polarity offset for each SET.

4.4.2 Kernel Function

The main operation of a single thread on the GPGPU is to compute the output SET at a single gate, given the SET at the gate inputs. Due to the SET parallelism, when processing a gate in each level, we need to propagate a SET group through a gate instead of a single SET. It is obvious that the computations for propagating each SET in a SET group for the same gate are independent of each other. Therefore, we can allocate these computations into multiple threads without incurring in any inter-dependency. On the other hand, propagating SETs of a SET group through a gate requires sharing some common data (pin-out delay, rising and falling delay, and signal probability). Based on this observation, the computation of a SET group for the same gate is assigned to the same thread block and the same multiprocessor. Having access to a fast on-chip shared memory, the shared data can be saved in the shared memory to reduce the memory access time.

When the target gates in all levels have been processed, one simulation step is completed and the control will return to the host CPU. The host reads the SET groups reached to the primary outputs, computes SER for the relevant gates, injects SET groups into the new site group, and invokes the next cycle.

4.5 Simulation Results

The proposed algorithm is implemented on CUDATM, enabling hardware from NVIDIA®. The host system for simulation contains Intel Xenon processors with two Processors and 64 GB RAM. The CUDA device is a NVIDIA Titan z quad-GPU-accelerator card with 4 × 5760 cores clocked at 876 MHz, each of which has exclusive access to 12 GB of global memory. We use 45 nm Nangate open cell library. We explore the efficiency of the proposed SER estimation method by applying the algorithm to ISCAS'85 and the large-scale EPFL benchmark circuits.

In the proposed method, each gate is injected with electrical charges of 32 levels, ranging from 30 to 200 fC. In order to provide an accurate and fair comparison, the runtime of the proposed method is compared with the same SER analysis approach, running in on the same CPU host system. In CPU-based method, after injecting the transient fault to each circuit gate, the SET is propagated through the fan-out cone of the fault site to reachable POs. The information related to benchmarks and the runtime derived using CPU-based and GPU-based methods are reported in Table 4.1. The first and second columns of Table 4.1 show the name of each circuit and the number of gates, respectively. Next two columns represent the number of primary inputs, primary outputs, and depth of each benchmark circuit. The fifth column shows the runtime of CPU-based SER estimation approach. In the sixth column, the runtime of the proposed method is reported and the last column represents the speedup of the proposed method for each circuit.

In the proposed method, the size of SET group is set to 32 (32 SETs with different pulse widths are simultaneously injected to each gate). Also, for each benchmark circuit, the conventional CPU-based SER estimation technique is repeated with 32 different SETs. For providing an accurate and fair comparison the pulse widths of injected SETs to each gate in both the techniques are similar. The simulation results indicate that the proposed GPU-based approach is much faster than conventional method by 920×, on average, when the SER analysis is performed for 32 different SETs. The least speedup is 360× which is achieved for C1908 benchmark, reaching up to 2390× for C6288. It should be noted that since the proposed framework employs and implements the three masking effects quite similar to the CPU-based SER estimation approach the accuracy of SER estimation would not be impressed.

Table 4.1 Runtime comparison

Circuit	#Node	#(PI,PO)	Level	CPU time (s)	GPU time (s)	Speedup
C432	160	36,7	30	136	0.08	1700×
C499	650	41,32	28	187	0.21	890×
C880	512	60,26	33	152	0.37	410×
C1355	653	41,32	30	204	0.53	384×
C1908	699	33,25	39	191	0.53	360×
C2670	756	233,144	38	210	0.55	381×
C3540	1467	50,22	52	812	0.84	966×
C5315	2115	178,123	41	1677	1.14	1471×
C6288	4507	32,32	122	8488	3.55	2390×
C7552	2534	207,108	60	2478	1.76	1407×
Bar	5k	135,128	12	6681.4	10.3	648×
Round	23k	256,129	87	11,056	17.8	621×
Log^2	54k	32,32	444	24,530	37.8	648×
Square	55k	64,128	250	25,919	43.5	595×

4.6   Conclusions

In this chapter, we presented a technique for reducing the SER estimation runtime by effectively exploiting the GPU resources. In the introduced method, multiple transient faults are injected simultaneously to several independent fault sites. Then, the transient faults are propagated through the circuit in a parallel approach in which the gates located in the same level are processed in parallel.

References

1. K. GULATI AND S. P. KHATRI, HARDWARE ACCELERATION OF EDA ALGORITHMS. SPRINGER, 2010.
2. E. SCHNEIDER, M. A. KOCHTE, S. HOLST, X. WEN, AND H.-J. WUNDERLICH, "GPU-ACCELERATED SIMULATION OF SMALL DELAY FAULTS," IEEE TRANS. COMPUT. DES. INTEGR. CIRCUITS SYST., VOL. 36, NO. 5, PP. 829–841, 2017.
3. S. ERCOLANI, M. FAVALLI, M. DAMIANI, P. OLIVO, AND B. RICCO, "ESTIMATE OF SIGNAL PROBABILITY IN COMBINATIONAL LOGIC NETWORKS," IN [1989] PROCEEDINGS OF THE 1ST EUROPEAN TEST CONFERENCE, 1989, PP. 132–138.
4. M. ZHANG AND N. R. SHANBHAG, "SOFT-ERROR-RATE-ANALYSIS (SERA) METHODOLOGY," COMPUT. DES. INTEGR. CIRCUITS SYST. IEEE TRANS., VOL. 25, NO. 10, PP. 2140–2155, 2006.
5. R. R. RAO, K. CHOPRA, D. T. BLAAUW, AND D. M. SYLVESTER, "COMPUTING THE SOFT ERROR RATE OF A COMBINATIONAL LOGIC CIRCUIT USING PARAMETERIZED DESCRIPTORS," IEEE TRANS. COMPUT. DES. INTEGR. CIRCUITS SYST., VOL. 26, NO. 3, PP. 468–479, 2007.
6. B. ZHANG, W. S. WANG, AND M. ORSHANSKY, "FASER: FAST ANALYSIS OF SOFT ERROR SUSCEPTIBILITY FOR CELL-BASED DESIGNS," IN PROC. INT. SYMP. QUAL. ELECTRON. DES. ISQED, PP. 755–760, 2006.
7. R. RAJARAMANT, J. S. KIM, N. VIJAYKRISHNAN, Y. XIE, AND M. J. IRWIN, "SEAT-LA: A SOFT ERROR ANALYSIS TOOL FOR COMBINATIONAL LOGIC," IN VLSI DESIGN, 2006. HELD JOINTLY WITH 5TH INTERNATIONAL CONFERENCE ON EMBEDDED SYSTEMS AND DESIGN, 19TH INTERNATIONAL CONFERENCE ON, 2006, P. 4–10.
8. J. M. BENEDETTO, P. H. EATON, D. G. MAVIS, M. GADLAGE, AND T. TURFLINGER, "VARIATION OF DIGITAL SET PULSE WIDTHS AND THE IMPLICATIONS FOR SINGLE EVENT HARDENING OF ADVANCED CMOS PROCESSES," IEEE TRANS. NUCL. SCI., VOL. 52, NO. 6, PP. 2114–2119, 2005.

Chapter 5
FPGA Hardware Acceleration of Soft Error Rate Estimation of Digital Circuits

Fault injection methods simulate the circuit behavior in the presence of faults and make the soft error rate (SER) estimation possible for a wide range of circuits. Based on fault injection approach, the fault injection methods are categorized as follows:

- Hardware fault injections
- Software-based fault injections
- Emulation-based fault injections

In hardware fault injection methods, a fabricated chip is bombarded with the ion or laser rays in order to inject faults and the effects of these faults are analyzed by running an application on the chip [1–3]. These methods are known as the fastest and also the most expensive ones among all of the fault injection methods. But the observability of the faults is at the lowest level in these methods.

In software-based fault injection methods, the behavior of the circuit which is described in software is simulated in the presence of injected faults and the SER is estimated by running several programs to track the effects of faults in the circuit [4–10]. In all these methods, the processing load is completely on CPU and because of the serial execution of instructions in said CPU, they are very time consuming. However, since the program is completely controlled by the user in software-based approaches, it provides a high observability of the internal values in the circuit.

In emulation-based methods [11–15], processes are loaded on an alternative hardware instead of CPU. Field-programmable gate array (FPGA) is a main candidate for this purpose because of its flexibility in different kinds of implementations. These methods usually have the most designing and implementation complexity among all of the fault injection methods, but since they are very fast, they are suitable for estimating the SER of very large circuits.

In this chapter, the probability propagation (software-based) approach and emulation-based method are combined in order to take the main advantages of both methods. The introduced approach benefits from fast runtime because of FPGA parallel processing capabilities besides the high observability as in software

© Springer Nature Switzerland AG 2021
B. Ghavami, M. Raji, *Soft Error Reliability of VLSI Circuits*,
https://doi.org/10.1007/978-3-030-51610-9_5

methods. In addition, regarding several tools developed, this approach has a high flexibility so that it can support a wide range of circuits based on the user needs and available resources. A novel algorithm is proposed for partitioning the circuit in order to simplify the calculation, but to carry out many computations of the said sub-circuits in parallel, by deploying the instrumentation hardware along with the sub-circuits in the FPGA. Using a fast approach based on a probability propagation method and utilizing an FPGA, the proposed approach is facilitated to estimate the SER of very large combinational circuits considering logical and electrical effects.

This chapter is organized as follows: in Sect. 5.1, the proposed framework is presented. Section 5.2 presents SER estimation using probability propagation approach. Section 5.3 explores the hardware implementation of the proposed framework. In Sect. 5.4, we explain the details about the implementation of the proposed approach on FPGA and the modifications needed to make it compatible with FPGA. Section 5.5 presents the experimental results and the acceleration obtained from the proposed approach. Finally, the chapter is concluded in Sect. 5.6.

5.1   The Introduced Framework

As shown in Fig. 5.1, the proposed approach consists of two main steps: (1) *instrumentation* and (2) *FPGA* run. In the instrumentation step, some modifications are done on the circuit in order to maximize the efficiency when running on FPGA. For

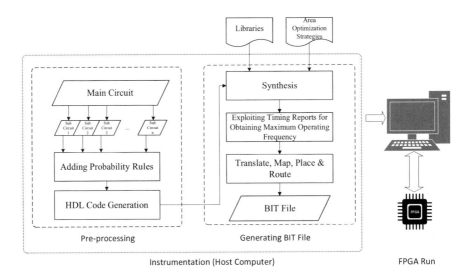

Fig. 5.1 An overview of the proposed approach composed of instrumentation phase (including the circuit partitioning, applying the probability rules, and generating the bit file for FPGA execution) and the FPGA run phase (including the execution of the proposed algorithm on FPGA and sending the results to the host computer)

this purpose, the circuit is first split into several independent sub-circuits such that they could fit on the FPGA. Then, a developed tool generates an HDL code of the sub-circuit based on a probability propagation model. The generated code will be used as an input for Xilinx commercial tools. In the last part of this step, a bit file containing the information about the circuit and the fault locations is downloaded on FPGA. In the second step, one or several sub-circuits (based on their sizes) are ran on FPGA to estimate the circuit SER. Since the proposed approach avoids unnecessary iterations and exploits FPGA for processing instead of CPU, it is expected that the introduced approach be much faster than the traditional ones.

5.2 SER Estimation Based on Probability Propagation Rules

In this section, we introduce the SER estimation using probability propagation approach.

For SER estimation, we use the fault probability propagation approach presented in [13]. In this approach, each signal in the circuit is modeled by a set of probabilities (i.e., the probabilities which show that the signal will have fault-free 1/0 and/or faulty 1/0). If no fault exists in the circuit, the probability of a signal to be faulty 1 or faulty 0 will be zero. On the other hand, if a fault occurs, all four probabilities may have nonzero values. A set of rules are defined to obtain the probabilities for the gate outputs based on their input values. Then, these probabilities are propagated through the gates until they reach the primary outputs (POs) of the circuits. By having the probabilities for the circuit POs, overall SER of the circuit can be computed. The overview of this approach is presented in Algorithm 5.1.

The main steps of this approach are as follows:

- Representing each signal of the gates by a set of four-value probabilities
- Injecting faults into the fault locations and observing the effects in the following paths
- Estimating the SER using the probabilities propagated to the POs

In more details, first the circuit is analyzed to identify the type and the number of gates in the circuit. Then, for each gate, the probability rules related to the type of the gate are applied on the inputs and outputs of the gates and this continues until the probability rules are applied on all gates of the circuit. By assigning fault-free initial values to the primary inputs (PIs), the probabilities are calculated for all gates and POs. In order to inject faults into the circuit, the probabilities for gate outputs obtained from the previous step are replaced by faulty probabilities. These faulty values propagate through the paths to POs. By comparing the faulty probabilities of the POs and the fault-free ones, the circuit SER can be estimated. In the following subsections, we explain more about the steps of the algorithm.

Algorithm 5.1 SER Estimation Using Probability Propagation Approach
1   dsnlst: Design Netlist
2   gatelist: Extract_Netlist_Adjacancy_List(dsnlst)
3   w: Pulse width
4   **for** each gate G_i in gatelist **do**
5 | List (G_i) = Extract and Sort Gates based on their type
6 | Apply_probability_rules (G_i);
7 | Compute_probabilities (G_i);
8   end
9   **for** each primary_input PI in List (G_i) **do**
10 | Apply_initial_values (PI);
11 end
12 **for** each primary_output PO in List (G_i) **do**
13 | Compute_Failure_Probabilty (PO);
14 end
15 List (G_i).Compute_SER();
16 /* Fault Injection */
17 Fault_List (G_i) = Select the gates to be targeted;
18 **for** each gate G_i in Fault_List (G_i) **do**
19 | Sub_List(G_i) = Extract and Sort on-path gates reachable from G_i;
20 | Event_List(G_i) = Add_Event $(P_a$, time=t);
21 | Event_List(G_i) = Add_Event $(P_{\bar{a}}$, time=t+w);
22 end
23 **for** each gate G_i in Sub_List(G_i) **do**
24 | Propagate_Events(G_i);
25 end
26 **for** each primary_output PO in Sub_List(G_i) **do**
27 | Compute_Failure_Probabilty(PO);
28 end
29 Sub_List(G_i).Compute_SER();

5.2.1   Four-Value Signal Probability Modelling

In the proposed signal probability model, each signal in the circuit can have the following four probabilities:

- 0: when a signal has fault-free probability of 0
- 1: when a signal has fault-free probability of 1
- 0_e: when a signal should have a probability of 1, but it has changed to 0 because of a particle strike
- 1_e: when a signal should have a probability of 0, but it has changed to 1 because of a particle strike

A set of four-value probabilities consisting of (P_1, P_0, $P_{\bar{a}}$, and P_a) is considered to describe the probability of each signal in the circuit. In this system, P_0 and P_1 represent the probability of being 0 and 1 at the output of a gate in fault-free condition, respectively. Also, other terms are needed to describe the probability of the transformed signal from its original value to the faulty one (transforming from 1 to 0 or from 0 to 1). The reason that we consider two terms for the faulty probability is that the transient pulse may have the same or different polarity from the original signal value. Hence, P_a represents the probability of a fault with the same polarity and ($P_{\bar{a}}$) represents the probability of a fault with the different polarity in the output of a gate in the presence of a fault caused by a particle strike.

Due to the type of each gate, a set of four-value probabilities should be calculated for its outputs based on its input probabilities. It should be noted that each gate has its individual set of equations developed to model the gate probabilities (i.e., AND gate equations differ from the NAND ones). For instance, a set of four-value probabilities are calculated for an n-input AND gate using its particular equations as follows [13]:

$$P_1(\text{OUT}) = \prod_{i=1}^{n} P_1(X_i) \tag{5.1}$$

$$P_a(\text{OUT}) = \prod_{i=1}^{n} \left[P_1(X_i) + P_a(X_i) \right] - P_1(\text{OUT}) \tag{5.2}$$

$$P_{\bar{a}}(\text{OUT}) = \prod_{i=1}^{n} \left[P_1(X_i) + P_{\bar{a}}(X_i) \right] - P_1(\text{OUT}) \tag{5.3}$$

$$P_0(\text{OUT}) = 1 - \left[P_1(\text{OUT}) + P_a(\text{OUT}) + P_{\bar{a}}(\text{OUT}) \right] \tag{5.4}$$

where X_i shows the input i of the gate and OUT represents its output.

By assigning probabilities of P_0 and P_1 to the PIs of the circuit, the probabilities are propagated through the paths to the POs of the circuit using the probability rules of (5.1)–(5.4) and the four-value probabilities are calculated for all of the gates. As an example, consider the circuit graph shown in Fig. 5.2. The probabilities for each

Fig. 5.2 A representative circuit graph

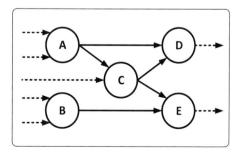

Table 5.1 The probabilities of the gate output calculation in only one traverse of the circuit

Gate	P_1	P_0	P_{1e}	P_{0e}
A	0.25	0.75	0	0
B	0.25	0.75	0	0
C	0.125	0.875	0	0
D	0.031	0.969	0	0
E	0.031	0.969	0	0

gate are computed by using the rules of (5.1)–(5.4) as represented in Table 5.1. After that the probabilities of PI (inputs of A, B, and C) being 0 or 1 are set to 0.5.

So, in the first pass, all the four-value probabilities are calculated for each output of the gates (fault-free condition). For each PO in the circuit, $P_{err}(PO_i)$ is defined as the probability for fault propagation to the PO_i in the circuit and is calculated as follows:

$$P_{err}\left(PO_i\right) = P_a\left(PO_i\right) + P_{\bar{a}}\left(PO_i\right) \tag{5.5}$$

where it is 0 in the fault-free condition.

5.2.2   Fault Injection Modeling

After calculating all the four-value probabilities for all the gates, the circuit is ready to inject faults. The fault model considered in this chapter is a bit-flip transient fault at the output of the gates. To perform the fault injection, the probabilities are exchanged based on the gate types. Probability exchanges are shown in Table 5.2 for a given gate. It should be noted that the probability exchange mechanism is the same for all gates and it does not depend on the gate type. For example, the probability of a faulty one at the gate output after fault injection will be the probability of a correct zero in the fault-free state; this resembles that a probability of zero is changed into a faulty one due to the particle strike.

After injecting a fault by changing the signal probabilities of the gate output, it should be propagated through the paths in the circuit until it reaches the POs. So, the PO probabilities are updated and the fault probability propagation will be involved in the PO probabilities.

5.2.3   SER Estimation

The circuit SER (SER_C) can be computed as

$$SER_C = \sum_{\forall g_i \in Circuit} SER_{g_i} \tag{5.6}$$

Table 5.2 Probability exchange mechanism for fault injection in the gate output

Original probability	Fault probability
$P_1(\text{OUT})$	$P_{0_e}(\text{OUT})$
$P_0(\text{OUT})$	$P_{1_e}(\text{OUT})$
$P_{1_e}(\text{OUT})$	$P_0(\text{OUT})$
$P_{0_e}(\text{OUT})$	$P_1(\text{OUT})$

where SER_{g_i} is the rate of soft error caused by an SET originated from gate g_i and being propagated to the circuit POs and it is calculated as follows:

$$\text{SER}_{g_i} = R_{\text{SEU}}\left(g_i\right) \times P_{\text{FP}}\left(g_i\right) \tag{5.7}$$

where $R_{\text{SEU}}(g_i)$ is the rate of transient fault that may occur in gate g_i which depends on many factors such as the particle energy level, size and type of the gate, and fabrication technology. Also, $P_{\text{FP}}(g_i)$ is the probability that a fault originated from gate g_i reaches one of the POs and is calculated as follows:

$$P_{\text{FP}}\left(g_i\right) = \left(1 - \prod_{j=1}^{k}\left[1 - P_{\text{err}}\left(\text{PO}_j\right)\right]\right) \tag{5.8}$$

where k is the number of the POs reachable from g_i.

So, in order to estimate the circuit SER after injecting a fault to each gate output, it is needed to perform this calculation only for the POs which have, at least, one reachable path from the faulty gate and not for all the POs. Remember that P_{err} was obtained by using Eq. (5.5), which uses Eqs. (5.1)–(5.4).

5.3   Hardware Implementation of the Probability Propagation

To have an effective implementation of the proposed SER estimation method, we present a novel design approach which takes advantage of both software- and emulation-based fault injection approaches. For this purpose, all of the computational processes are performed on FPGA to obtain high acceleration by using its parallel processing capabilities and on the other hand the system condition is observable after each change to the circuit, which offers high observability to the designer. Hence, the proposed probability propagation model should have enough compatibility to be implemented on FPGA at a parallel manner. So, each gate in the circuit is converted into a computational block to run on FPGA. Then, these blocks are modified by fault injection and circuit SER is calculated by analyzing the results. This will be applicable in three steps:

- Generating computational blocks for the probability propagation model
- Fault injection
- SER estimation

More details about the mentioned steps are provided in the following subsections.

5.3.1   Generating Computational Blocks for the Probability Propagation Model

In order to implement the proposed approach on FPGA, each gate should be converted into a hardware computational block based on its type. The objective of this block is to apply the probability propagation rules on signals (Fig. 5.3). As mentioned earlier, in the proposed model, each signal could have one of the probabilities of 0, 1, 0_e, and 1_e based on being whether in the fault-free or in the faulty condition. Hence, for each block, all input or output signals are expanded into four signals to represent the four-value probabilities (P_1, P_0, P_{1_e}, and P_{0_e}).

After expanding the gate signals to the four-value probabilities, the output probabilities are calculated for each gate based on PI initial probabilities and the gate

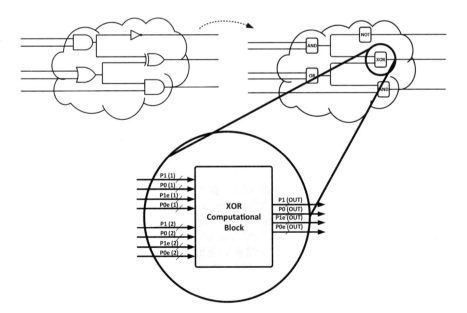

Fig. 5.3 Conversion of a logical gate into a computational block for probability propagation model

types, as described in Sect. 5.2. For this purpose, a methodology is developed which takes the HDL code of the circuit as an input and identifies all elements in the circuit by parsing the code in order to categorize the gates based on their type. Then, by referring to a modified library, the traditional model of the gate is replaced with its signal probability model based on the gate type. In addition, some extra hardware is added to the output of the gate to make the fault injection possible. Finally, all connections are rebuilt based on the circuit initial HDL code and a new one is generated as the output of the program (Fig. 5.4). The generated output file is fully compatible to be synthesized with the wide range of synthesis frameworks making the proposed approach flexible and not being limited to a specific technology.

5.3.2   Fault Injection

In order to inject faults into a given gate in the circuit, the signal probabilities need to be changed in the target gate. Changing the probabilities in a gate depends on the fault model being used. In order to simulate the single-event transient fault model on the output of a gate, the probabilities of the output signal are changed as described in Sect. 5.2. This change lasts for a specific period of time called "fault time" which represents the duration of presence of the transient pulse in the circuit. Meanwhile, any desired value of the signals could be captured easily to keep the observability of the proposed approach at an acceptable level.

After modeling the circuit by computational blocks, an extra part is required to be added to each block for injecting faults. For this purpose, a control unit is added to each gate output to simulate the impact of a transient fault in the circuit by changing the probabilities. As shown in Fig. 5.5, if Fa_en signal is triggered in the control unit, the control unit acts as a switch and changes the probabilities due to the signal change mechanism shown in Table 5.2. On the other hand, if Fa_en signal is not triggered, the control unit acts as a buffer and only passes the probabilities. This

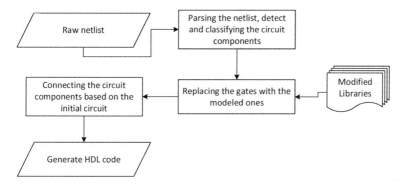

Fig. 5.4 The HDL code generation flow in the proposed developed tool to apply the probability propagation rules

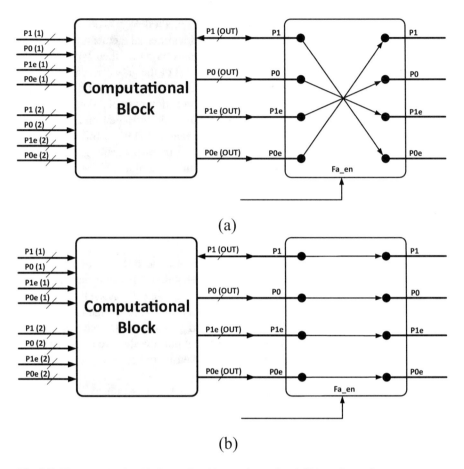

Fig. 5.5 The control unit with Fa_en signal is (**a**) triggered and (**b**) not triggered

mechanism allows the fault injection to be performed immediately at any desired location in the circuit without needing to wait for repetitions of injecting faults in previous gates since the signals keep their values. On the contrary, to inject a fault at a gate in CPU-based probability propagation approach, a time is needed to produce the probabilities from the previous gates (propagation time) and this time varies based on the distance of the gate from the PIs.

5.3.3   SER Estimation

After creating the probability propagation model with capability of fault injection at the output of the gates, it has to be run on FPGA. So, some modifications have to be done on the generated synthesizable HDL code to be ready for running on FPGA. For generating the bit file, the synthesis process starts to choose the proper hardware blocks in order to utilize the FPGA resources efficiently. After the synthesis process, information about the required amount of resources and timing analysis will be available [15]. In the next step, the synthesized file is analyzed to allocate the FPGA resources to the different circuit components. Finally, after routing and meeting the timing constraints, the executable file for running on the FPGA is generated.

As mentioned before, it is impossible to model the probabilities with a single-logical-value signal since it can hold only two values of logical one and logical zero. Hence, it is needed to expand the signals for carrying a probability between 0 and 1. For this purpose, a fixed-point signal is considered to represent each of the four-value probabilities (P_1, P_0, P_{1_e}, and P_{0_e}). So, to have all the probabilities together, four fixed-point signals are needed (Fig. 5.6).

After injecting faults to each fault location in the circuit, the probabilities are calculated for all of the POs in the circuit and they are sent to the host computer. By having the fault probabilities for the POs (P_a, $P_{\bar{a}}$), P_{err} can be calculated for all of the POs in the circuit using Eq. (5.5). After that, with P_{err} available for each PO and using Eqs. (5.6)–(5.8), the failure probability of the circuit and then the circuit SER are estimated.

5.4 Efficient Implementation with FPGA

5.4.1 Area Problem

In the previous section, the hardware implementation of the probability propagation approach is explained. However, in the path to implement the mentioned approach, there is a main challenge which is the significant area overhead caused by the probability propagation model. In this model, each signal is converted to four probability

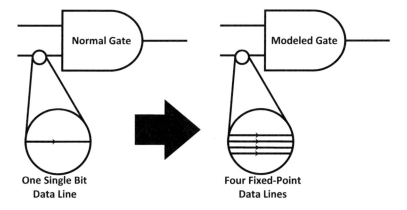

Fig. 5.6 Converting a single-logical-value signal in the normal condition to four-fixed-point-value probability in the proposed modeled gate

values (Fig. 5.6). Hence, significant amount of area overhead will be imposed compared to the normal condition in the probability propagation model. This overhead is so high that it makes the implementation of the proposed approach impossible for most of the available FPGAs in the market. To encounter the mentioned challenge, the circuit is split into smaller partitions to meet the hardware resource needs. However, by splitting the circuit into the smaller sub-circuits, a new challenge appears; that is, a complete version containing the whole circuit components should be available to make the circuit SER estimation possible. To overcome this problem, the approach presented for the circuit partitioning splits the circuit in a way that it keeps the independency of each sub-circuit to make the calculation of the failure probability for each one possible without needing the data of other sub-circuits. Then, some timing and area modifications are done in order to maximize the efficient utilization of the FPGA resources.

These modifications include using the maximum area reduction strategies to optimize the resource utilization on FPGA. These strategies are modified such that the design is completely customized considering the available FPGA. For instance, during the synthesis process, all tristate elements are replaced with their logical equivalents to make the design executable on FPGAs without the internal tristate elements. This imposes a lot of area overhead to the design. As the FPGA used in this experiment supports the internal tristate features, we disable the tristate replacement to avoid such area overhead. Also, since the proposed approach is specialized for the combinational circuits, the clock enable (CE) signal allocation is avoided for a large number of flip-flops in the circuit to prevent consuming a large amount of resources used for routing the CE signal.

5.4.2   Circuit Partitioning

In the probability propagation approach, the circuit SER is estimated using the circuit failure rate. As explained in Sect. 5.2, calculating the fault probability propagation for a gate only depends on the POs reachable from that gate. Hereinafter, we call the set of the POs reachable from a gate and all gates located on the paths between the gate and those POs as the "output cone" of the gate (Fig. 5.7). So, in order to obtain the circuit failure rate for the injected fault in a gate, the fault should be propagated through all gates in the gate output cone. However, partitioning the circuit only based on the output cones of the gates leads to creating a large number of sub-circuits and thus is not an efficient choice. On the other hand, the size of some sub-circuits for the gates close to the PIs will be near to the main circuit and the partitioning will be actually useless. Hence, it is required to present a partitioning approach which offers an acceptable trade-off between the size and number of sub-circuits.

Since the calculation of the failure probability propagation to a PO is independent from other POs, circuit partitioning in a way that each sub-circuit has only one PO could be an appropriate alternative to handle the said trade-off between the size

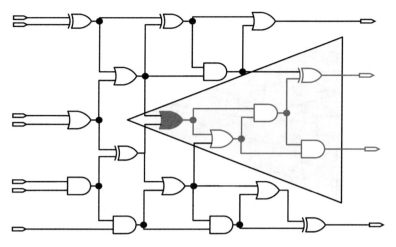

Fig. 5.7   An example of a gate in the circuit with its output cone

and the number of the sub-circuits. However, it is possible to have more than one PO in a gate output cone (Fig. 5.7). As mentioned before, in order to compute the probability of a failure originated from injecting a fault in a given gate, it is required to have the gate and all the gates in its output cone. So, the gates that have a reachable path to more than one PO (several POs on their output cone) appear in several sub-circuits to make sure that the relation of a gate and all of the POs reachable from it is kept. As shown in Fig. 5.8, gate 3 has paths to both POs 12 and 13 and hence it appears in two sub-circuits.

To partition the circuit based on the POs, we first choose a PO and then include all the gates that have at least one path to that PO; this set is considered as a sub-circuit. Then, this process is repeated for the rest of the POs to build all the sub-circuits. Extracting the sub-circuits from the main circuit is done in the following steps:

1. *Building the main graph*: The main circuit graph (*G*) is constructed such that each node (*N*) represents the gates in the circuit and each edge (*E*) shows the connection between two directly connected gates.
2. *Circuit levelization:* The level of each node is obtained by using a modified linear topological sorting algorithm. In a traditional topological sorting algorithm, for each connection from vertex *u* to vertex *v*, *u* comes before *v* but, in such a condition in our ordering algorithm, vertex *v* comes before vertex *u*. So, all POs in the circuit are in the first level (0) since they are met first in our sorting algorithm and they form the primary output list (POL).
3. *Tagging the nodes:* With reverse traversing from a PO to the PIs, all visited nodes (*n*) will get a tag (*T*). In this step, we assign an individual tag to each PO and continue assigning the said tag to all nodes during the path from PO to PI(s). So, each node will have the same tag as the PO tag until the end of the reverse traversing from a PO to PI(s). The mutual tag for two nodes means that there is at

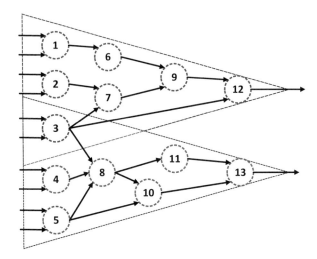

Fig. 5.8 The main circuit and sub-circuits after partitioning the main circuit based on POs

least one path between those two nodes in the circuit. So, a list of tags for each node ($TL(n)$) is obtained using the following equation:

$$TL(n) = \bigcup T(n)$$

(5.9)

where $T(n)$ is the assigned node tag in each traverse.

4. *Extracting the input cones based on each PO:* Until this step, each node has one or several tags obtained from the previous step. In order to extract the input cones of each PO, it is needed to identify the nodes that have at least one path to the said PO. As mentioned before, existing a path between two nodes means that they have a mutual tag in their tag list. So, performing a comparison between the PO tag and the ones in tag list of each node can define the input cone for the said PO. To do that, one of the POs from the POL is chosen at first. Then, the PO tag obtained from step 3 is compared with the tag list of the nodes in the circuit. If a similarity between PO tag and any of the tags belonging to the tag list of a node ($TL(n)$) is found, that gate is considered as a member of sub-circuit. It should be noted that if a node has more than one tag, it shows that it belongs to several sub-circuits (Fig. 5.8).

5. *Building the sub-circuit:* Up to this point, we have the required information about the nodes located in the input cone of the POs. However, we also have to know the position of each node belonging to an input cone to build a sub-circuit. This is done by merging the data obtained from step 2 and step 3; each sub-circuit containing the PO node i (Sub_i) is built based on the following equation:

$$\forall i \in POL \rightarrow Sub_i = PO_i$$
$$\bigcup_{n \in G} \left\{ n | \left(TL(n) = TL(T(n)) \right) \text{ and } \left(TL(n) = TL(PO_j) \right) \right\}$$

(5.10)

where PO_j represents the sub-circuit related to PO j and PO(N).

The generated sub-circuits have the following features:

(a) Each sub-circuit has only one PO.
(b) The number of sub-circuits is the same as the number of POs in the main circuit.
(c) Sub-circuits are completely independent (to calculate PO probabilities in a sub-circuit, only the gates located in the sub-circuit are needed).

Using the mentioned algorithm for partitioning the circuit, probabilities for each PO can be calculated independently. This makes the parallel computations for several sub-circuits possible considering their size. The size of each sub-circuit can vary from two gates to all gates in the main circuit except the POs. This size is proportional to the number of the reconvergent paths in the circuit. By having the probabilities for a PO after injecting a fault into the gate (g_i), P_{err} can be computed using Eq. (5.5). It should be noted that if a gate appears in several sub-circuits, it will cause several POs in different sub-circuits to have several probabilities of P_{err}, and all these probabilities should be considered in Eq. (5.8) in order to compute the $P_{FP}(g_i)$. As the final calculation is done on the host computer to compute the SER of the circuit, the presence of a gate in different sub-circuits is checked by referring to the tag list of each gate. For instance, assume the computation of P_{FP} for an injected fault into gate 3 in Fig. 5.8. As shown, gate 3 belongs to two different sub-circuits. Hence, for computing $P_{FP}(g_3)$, both probabilities of P_{err} for PO12 and PO13 are needed to be considered in Eq. (5.8). The host computer first checks the tag list of the gate 3 and then refers to the set of results obtained from the FPGA run. After that, it picks the appropriate probabilities and inserts them in Eq. (5.8) and this procedure continues until all P_{FP} probabilities are computed. After computing the P_{FP} probabilities for the gates using Eq. (5.8), the SER of the gates is computed using Eq. (5.7) and consequently the overall SER of the circuit can be obtained by Eq. (5.6).

5.5 Experimental Results

This section presents the experimental results obtained from the parallel execution of the proposed approach on FPGA and the amount of acceleration obtained in comparison with the execution of the same algorithm on CPU. A set of experiments called CPU execution are run on a computer with Intel Core i7 720 processor equipped with 6 GB DDR3 RAM while in another set of experiments called FPGA execution, a Xilinx KC-705 board with XC7K325T FPGA is used. In these experiments, ISCAS'85 benchmark circuits are used. In the instrumentation phase, to generate a synthesizable file, we use our developed tool that makes the HDL code suitable for the input of Xilinx commercial tools. Then, by using the Xilinx tools and required possible modification in different parts of these tools, an executable file will be generated to be downloaded on FPGA.

5.5.1 Resource Utilization

In this part, we investigate the hardware overhead imposed to the circuit during the instrumentation phase. Since in the proposed approach, our goal is to achieve the maximum acceleration, we choose the instrumentation-based approach for injecting the faults. The instrumented block is composed of a computational block which generates the probabilities based on the type of each gate in the circuit and a control unit which is responsible for injecting the faults at the output of the gates.

Table 5.3 shows the overall FPGA utilization for implementing the sub-circuits to perform the proposed approach. Columns 2 and 3 show the number of the gates and the number of sub-circuits after the partitioning process, respectively. The number and the percentage of Slice LUT utilization are brought in columns 4 and 5 of this table. The mentioned resource utilization belongs to the whole computational block (Fig. 5.5).

Also, column 6 shows the minimum number of FPGA configurations required for SER estimation of a benchmark. It should be noted that if several sub-circuits can be fitted on the FPGA, we bring them all on the FPGA and perform the parallel calculations for each sub-circuit independently. Hence, the information reported in Table 5.3 does not necessarily belong to one sub-circuit, but the approximate logic needed for each sub-circuit can be reached by dividing the overall utilization to the number of the sub-circuits.

Table 5.3 Hardware utilization for implementation of different sub-circuits on the FPGA

Benchmark	#Gates	Number of sub-circuits	Number of slice LUTs	Slice LUTs used (%)	Number of FPGA configurations
C432	160	7	17,887	17.6	1
C499	202	18	93,026	91.7	1
C880	383	16	97,421	96.1	1
C1355	546	12	92,325	91	1
C1908	880	8	90,430	89.1	2
C2670	1193	121	100,051	98.7	3
C3540	1669	7	91,473	90.2	2
C5315	2307	52	98,022	96.7	2
C6288	2416	4	96,583	95.2	3
C7552	3512	95	99,782	98.4	3

5.5.2 Circuit Partitioning Acceleration Results

In this part, the impact of the proposed circuit partitioning on the runtime of the implemented approach is investigated. As mentioned before, the overall process for estimating the SER includes two main phases:

1. Obtaining the P_{err} for all of the primary outputs
2. Estimating the circuit SER using Eqs. (5.8), (5.7), and (5.6)

The second phase is performed on the host computer and is exactly the same in both software-based and emulation-based approaches. So, the runtimes for the CPU and the FPGA only include the time needed for completing the first phase. Hence, for a fair comparison between the CPU and the FPGA runtime, we only considered the time that $P_{err}(PO_i)$ is obtained for all the primary outputs of the circuit based on Eq. (5.5).

Table 5.4 shows the results obtained from implementing the proposed approach before and after applying the proposed partitioning approach on the benchmarks. The first column includes the benchmark name. The second part of the table (columns 2, 3, and 4) represents the runtime of executing the proposed approach on CPU and the parallel execution on FPGA before and after partitioning the circuits. As shown in this part, implementing the proposed approach is not possible for large circuits and even some of the midrange-sized ones without circuit partitioning since they cannot be fitted on the FPGA. However, the FPGA execution runtime after partitioning the circuits is highly dependent on the size of the largest sub-circuit. The last part of the table is dedicated to represent the amount of acceleration obtained from the FPGA execution compared to the CPU execution before and after the circuit partitioning. The results show that the probability propagation approach can be implemented on the whole ISCAS'85 benchmarks after the circuit partitioning. Moreover, the lowest speedup belongs to C6288 benchmark as it has the largest

Table 5.4 Runtime of the execution of the proposed approach of the main circuits on FPGA and CPU and the obtained speedup due to the proposed partitioning approach

Benchmark info.	Runtimes (in s)		Speedup		
Name	CPU	FPGA	FPGA (partitioned)	FPGA	FPGA (partitioned)
C432	2.17	5.432E−05	4.053E−05	3.99E+04	5.35E+04
C499	25.48	3.465E−05	1.482E−05	7.35E+05	1.72E+06
C880	1.90	1.129E−04	2.574E−05	1.68E+04	7.38E+04
C1355	3.81	1.622E−04	4.347E−05	2.35E+04	8.76E+04
C1908	10.22	NA	1.052E−04	–	9.71E+04
C2670	18.59	NA	2.355E−04	–	7.89E+04
C3540	50.36	NA	6.421E−04	–	7.84E+04
C5315	126.44	NA	2.617E−04	–	4.83E+05
C6288	153.49	NA	2.438E−03	–	6.30E+04
C7552	210.08	NA	1.345E−04	–	1.56E+06

number of reconvergent paths that causes the generated sub-circuits not to be balanced enough.

5.6 Conclusion

In this chapter, an efficient approach for the circuit SER estimation is developed based on parallel execution capability in FPGAs. For this purpose, a probability propagation approach is used along with several modified tools to accelerate the time needed for SER estimation of digital circuits. Also, an algorithm is presented for partitioning the circuits that enables the required parallelism for a wide range of circuits by implementing on FPGAs.

References

1. S.Z. Shazli, M.B. Tahoori, "Using Boolean Satisfiability for Computing Soft Error Rates in Early Design Stages," Microelectronics Reliability, Vol. 50, No. 1, pp. 149–159, 2010.
2. N. Kehl and W. Rosenstiel, "An efficient SER estimation method for combinational circuits," IEEE Trans. Reliability, vol. 60, no. 4, pp. 742–747, 2011.
3. L. Chen, M. Ebrahimi, M. B. Tahoori, "CEP: Correlated Error Propagation for Hierarchical Soft Error Analysis," Journal of Electronic Testing (JETTA) Vol. 29, pp. 143–158, 2013.
4. G. Asadi, M.B. Tahoori, "An analytical approach for soft error rate estimation in digital circuits," in Proc. IEEE International Symposium on Circuits and Systems (ISCAS), Vol. 3, pp. 2991–2994, 2005.
5. L. Entrena, M. G. Valderas, R. F. Cardenal, M. P. Garcia, and C. L. Ongil, "SET emulation considering electrical masking effects," IEEE Trans. Nuclear Science, vol. 56, pp. 2021–2025, 2009.
6. R. Nyberg, J. Heyszl, D. Heinz, and G. Sigl, "Enhancing fault emulation of transient faults by separating combinational and sequential fault propagation," in ACM Great Lakes Symposium on VLSI, pp. 209–214, 2016.
7. L. Entrena, M. Valderas, R. Cardenal, A. Lindoso, M. P. Garcia, C. Lopez-Ongil "Soft error sensitivity evaluation of microprocessors by multilevel emulation-based fault injection," IEEE Trans. Computers (TC), Vol. 61, No. 3, pp. 313–322, 2012.
8. Anees Ullah, Pedro Reviriego, Juan Antonio Maestro, "An Efficient Methodology for On-Chip SEU Injection in Flip-Flops for Xilinx FPGAs", IEEE Trans. Nuclear Science, vol. 65, no. 4, pp. 989–996, 2018.
9. F. Serrano, J. A. Clemente, H. Mecha, "A Methodology to Emulate Single Event Upsets in Flip-Flops Using FPGAs through Partial Reconfiguration and Instrumentation", IEEE Trans. Nuclear Science, vol. 62, pp. 1617–1624, 2015.
10. M. Ebrahimi, A. Mohammadi, A. Ejlali, and S. G. Miremadi, "A fast, flexible, and easy-to-develop FPGA-based fault injection technique," Microelectronics Reliability, vol. 54, no. 5, pp. 1000–1008, 2014.
11. H. Quinn, D. Black, W. Robinson, S. Buchner, "Fault Simulation and Emulation Tools to Augment Radiation-Hardness Assurance Testing", IEEE Trans. Nuclear Science, vol. 30, no. 3, pp. 2119–2142, 2013.

12. P. SHIVAKUMAR, M. KISTLER, S. W. KECKLER, D. BURGER, AND L. ALVISI, "MODELING THE EFFECT OF TECHNOLOGY TRENDS ON THE SER OF COMBINATIONAL LOGIC," IN PROC. INT. CONFERENCE ON DEPENDABLE SYSTEMS AND NETWORKS (DSN), JUN. 2002, PP. 389–398.
13. G. ASADI AND M. B. TAHOORI, "AN ACCURATE SER ESTIMATION METHOD BASED ON PROPAGATION PROBABILITY," IN PROC. DESIGN AUTOMATION AND TEST IN EUROPE CONFERENCE (DATE), PP. 306–307, 2005.
14. XST USER GUIDE FOR VIRTEX-6, SPARTAN-6, AND 7 SERIES DEVICES, [ONLINE]. AVAILABLE: https://www.xilinx.com/support.
15. TIMING CLOSURE USER GUIDE, [ONLINE]. AVAILABLE: https://www.xilinx.com/support.

Chapter 6
Soft Error Tolerant Circuit Design Using Partitioning-Based Gate Sizing

Gate sizing is one of the simplest yet effective methods of soft error hardening techniques. In this technique, the sizing of a gate (i.e., *W/L* in gate transistors) is perturbed to increase its resilience against the particle strikes [1–3]. In the gate sizing technique, the gates may be upsized in order to increase the output capacitances of the gates in which charging/discharging results in a single-event transient (SET) after a strike or downsized to increase the electrical attenuation occurred on the SET while it is propagating through the gates in the circuit.

This chapter presents a fast and efficient technique for resizing the large-scale combinational circuits [4]. In this methodology, the circuit is partitioned into a set of smaller sub-circuits based on the cone structures which are originated from the primary outputs (POs). The extracted sub-circuits are topologically levelized and then, by starting from the minimum level, the sub-circuits located at the same level are resized individually and independently. This procedure is continued level by level until the sub-circuits in all levels are resized. During resizing process, after each gate sizing, the sub-circuit error probability (SEP) of the sub-circuit in which the resized gate is located is computed. Based on the computed SEP, the effects of gate sizing on the total circuit SER are evaluated. Therefore, instead of recomputing the circuit soft error rate (SER) for each gate resizing that is far too time consuming, it only requires to compute the SEP of the sub-circuits in which the resized gates are located. Since the SEP computation is much faster than circuit SER computation, evaluating the effects of gate sizing on circuit SER locally by the SEPs results in significant acceleration in the gate sizing optimization algorithm.

This chapter is organized as follows. Section 6.1 describes the gate sizing algorithm. Section 6.2 presents the results of the proposed technique. Finally, the chapter concludes in Sect. 6.3.

© Springer Nature Switzerland AG 2021
B. Ghavami, M. Raji, *Soft Error Reliability of VLSI Circuits*,
https://doi.org/10.1007/978-3-030-51610-9_6

6.1 Gate Sizing Technique

In all gate sizing techniques, after resizing each gate, the effect of the resized gate on the circuit SER is investigated [1–4]. Hence, one of the most important steps in gate sizing techniques is investigating the effect of gate sizing on circuit SER. From the SER evaluation procedure standpoint, the gate sizing technique is classified into the backward and forward propagation approaches. Forward propagation-based techniques include the ones in which after resizing a gate, the SER of the parts of the circuit that have been affected by the resized gate is recomputed [1, 5, 6]. To this end, the SER of the gates, which are located in the fan-out cone of the resized gate, is recomputed (Fig. 6.1a). The second group of gate sizing techniques (i.e., backward propagation-based techniques) uses the vulnerability concept instead of circuit

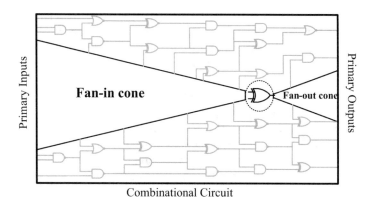

(a) fan-in and fan-out cone of a gate close to POs

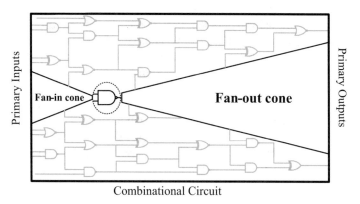

(b) fan-in and fan-out cone of a gate close to PIs

Fig. 6.1 Fan-in and fan-out cone of a gates in a combinational circuit. (**a**) Fan-in and fan-out cone of a gate close to POs. (**b**) Fan-in and fan-out cone of a gate close to PIs

SER in order to evaluate the circuit reliability against the soft errors [7]. Circuit gate vulnerability is calculated by propagating vulnerability parameters backwardly from POs to PIs [8]. Although, using the vulnerability analysis, we do not need to recompute the circuit SER for each resizing action, but the vulnerability parameters of the gates located in the fan-in cone of the resized gate are changed after resizing it. Hence, by each gate sizing action, the vulnerability of the gates in the fan-in cone of the resized gate should be updated (Fig. 6.1a).

It this section, by partitioning the circuit into a set of smaller sub-circuits, we can combine the two approaches (i.e., SER analysis (forward propagation-based technique) and vulnerability analysis (backward propagation-based technique)) to resize the gates of large-scale combinational circuits much faster in order to minimize their SER. We introduce a soft error-tolerant combinational circuit design technique which is based on minimizing the probability of transient fault generation and propagation through the circuit gates. In this technique, the circuit is partitioned into a set of small sub-circuits and then the probability of generation and propagation of transient faults in each sub-circuit, which is named as SEP (sub-circuit error probability), is minimized using a genetic gate sizing optimization algorithm on the sub-circuits. Reducing the sub-circuits SEP leads to reducing the total soft error rate of the original circuit.

6.1.1 Circuit Partitioning

The original circuit is split into a set of small sub-circuits using cone structures which are originated from circuit POs. The parts of the circuit which are located only in one cone are considered as the circuit clusters. In addition, the parts of the circuit which have been shared between several cones are also considered as clusters. After determining the POs and PIs of clusters, each cluster is defined as a sub-circuit. Cone structure is introduced as a set of gates which are located between a specific PO and PIs. Figure 6.2 shows the graph representation of a combinational circuit and its cone structures.

A combinational circuit is partitioned into a set of small sub-circuits by the following six steps:

1. *Circuit graph (CG) construction*: All circuit nodes from PO to every reachable PI are included in a tree while the POs are its roots and a subset of PI is its leaves. This tree is extracted using the graph backward traversing algorithm.
2. *Levelizing*: The circuit gates are levelized using a topological sorting algorithm. It is notable that the traditional topological sorting is performed with a modification on the edge direction of the circuit graph; that is, if there is an edge (u, v) in the circuit graph, u appears after v in the list.
3. *Tagging*: We start to backwardly traverse the CG from the roots of circuit tree. During each backward traversing, a tag T (corresponding to the root number) is assigned to the nodes which are visited. Finally, each graph node has a list of

Fig. 6.2 A graph
representation of a
combinational circuit and
its cone structures

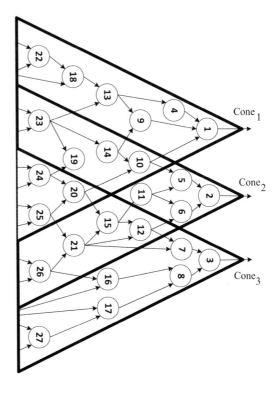

tags which is called as tag list (TL). So, for each node n in the CG, we have a tag
list TL(n) which is defined as follows:

$$TL(N) = \bigcup_{Pa(N)} TL(Pa(N))$$

(6.1)

where Pa(N) is the parents of node n in the CG.

4. *Cluster root extracting*: A graph cluster is defined as the set of nodes with the
   same TL. In a cluster, the node which has the minimum topological level is
   named as cluster root (CR). In order to extract the graph clusters, a list containing all the CRs is constructed as follows (we name this list as clusters roots list
   (CRL)): at first, the roots of circuit tree are added to the CRL. Then, the nodes
   whose parents' TL are different from their TL are added to the CRL; that is,
   these nodes are the root of the clusters that are located between two or more
   overlapping cones. The overlapping regions in a CG are illustrated in Fig. 6.2.
   The degree of shading in a region reflects the amount of overlap between cones.
5. *Clustering*: A cluster originating from node i (Cluster$_i$) in the graph is defined as
   follows:

$$\forall i \in \text{CRL} \rightarrow \text{Cluster}_i = \text{CRL}_i \bigcup_{\forall N \in \text{CG}} \{ N\# \ (\exists \text{Pa}(N)[\text{TL}(N)$$

$$= \text{TL}(\text{Pa}(N))]) \text{ and } (\text{TL}(N) = \text{TL}(\text{CRL}_i))\} \tag{6.2}$$

where CRL_i is a cluster root.

6. *PO and PI assignment:* The nodes in the CG which are connected to a circuit POs are considered as the virtual primary output (VPO). In addition, the nodes which at least have a parent with a different TL from its own TL are considered as a VPO:

$$\forall N \in \text{CG} \rightarrow \text{VPO}_i = N\# \ (N \in \text{CRL}) \text{ or } (\exists \text{Pa}(N) | \text{TL}(\text{Pa}(N) \neq \text{TL}(N))) \tag{6.3}$$

The node inputs which are connected to one of the circuit PIs or originated from the child nodes with different TLs are considered as virtual primary inputs (VPI). For these inputs, we have

$$\forall E \in \text{CG} \rightarrow \text{VPI}_i$$
$$= E \Big| \big[(E \in \text{PI}_{\text{circuit}}) \big] \text{ or } \big[(\text{TL}(u) \neq \text{TL}(v)) \big] \Big|$$
node v is connected to node u by edge E
$$\tag{6.4}$$

where E is an edge in the graph and $\text{PI}_{\text{circuit}}$ shows the circuit PIs.

The original circuit is partitioned into a set of clusters using the proposed cone structure approach. By assigning the VPIs and VPOs to each cluster, the circuit is split into a number of independent smaller sub-circuits. Figure 6.3 shows the partitioned graph of the graph in Fig. 6.2.

It is notable that such cone-oriented circuit partitioning facilitates levelizing the extracted sub-circuits which in turn prevents excessive recomputation of sub-circuit vulnerability during gate sizing (it will be discussed in more detail in the following sections).

6.1.2   Sub-circuit Error Probability Computation

The contribution of a sub-circuit in the total circuit SER includes the effect of sub-circuit in the generation of transient faults and their propagation. These effects are a function of the sub-circuit structure and their location in the main circuit. To evaluate the effects of the sub-circuit SC_i in the generation and propagation of transient faults in the main circuit, sub-circuit error probability (SEP_{SC_i}) parameter is introduced. This parameter is defined as the sum of the error latching probability (ELP) on the circuit POs for all gates in the sub-circuit SC_i, i.e.,

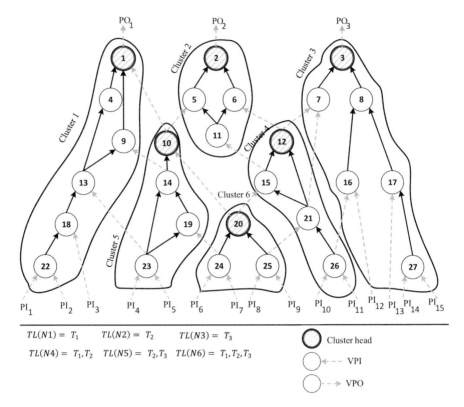

Fig. 6.3 The partitioned graph of the graph in Fig. 6.2

$$\mathrm{SEP}_{\mathrm{SC}_i} = \sum_{j \in N_{\mathrm{node}}(\mathrm{SC}_i)} \mathrm{ELP}^j_{\mathrm{SC}_i} \qquad (6.5)$$

where $N_{\mathrm{node}}(\mathrm{SC}_i)$ is the union of the sub-circuit SC_i inputs and the nodes in the sub-circuit SC_i which may be hit by radiation-induced particles. So, for N_{node} in Eq. (6.5), we have

$$N_{\mathrm{node}}\left(\mathrm{SC}_i\right) = \mathrm{gates}_{\mathrm{SC}_i} \cup \mathrm{VPI}_{\mathrm{SC}_i} \qquad (6.6)$$

where $\mathrm{VPI}_{\mathrm{SC}_i}$ is the virtual primary inputs of sub-circuit SC_i and $\mathrm{gates}_{\mathrm{SC}_i}$ shows the gates located in sub-circuit SC_i. In fact, in order to model the transient faults which are originated from the other sub-circuits and propagated from sub-circuit VPIs to sub-circuit VPOs (i.e., propagation effect of the sub-circuit), we consider the VPIs as the same as the other circuit nodes which generate transient faults in their output.

In Eq. (6.5), $\mathrm{ELP}^j_{\mathrm{SC}_i}$ is error latching probability for node j in the sub-circuit SC_i and defined as the product of SET generation probability at node j ($P_{\mathrm{gen}}(j)$), the probability of SET propagating from node j to the sub-circuit VPOs ($P_{\mathrm{prop}}(\cdot)$), and the probability of SET propagating from sub-circuit VPOs to the circuit POs and latching in the circuit flip-flops ($P_{\text{Latching}}(\cdot)$):

$$\text{ELP}_{\text{SC}_i}^{j} = \sum_{\forall k \in \text{VPO}_{\text{SC}_i}} P_{\text{gen}}\left(j\right) * P_{\text{prop}}\left(j,k\right) * P_{\text{Latching}}\left(k\right) \tag{6.7}$$

where $P_{\text{prop}}(j,k)$ shows the probability of generated SET in the node j propagating to the VPO k of sub-circuit SC_i. $P_{\text{prop}}(j,k)$ is dependent on the electrical and logical masking of the gates located in the SET propagation path and is computed as follows:

$$P_{\text{prop}}\left(j,k\right) = \sum_{k \in \text{VPO}_{\text{SC}_i}} P_{\text{elec}}\left(j,k\right) * P_{\text{logic}}\left(j,k\right) \tag{6.8}$$

where VPO_{SC_i} is the set of SC_i sub-circuit VPOs, $P_{\text{logic}}(j,k)$ denotes the sum of logical probabilities of transient faults propagating through the path between node j and VPO k, and $P_{\text{elec}}(j,k)$ is the electrical probability, representing the electrical masking of the gates located in the propagation path of transient fault in the sub-circuit SC_i.

In Eq. (6.7), $P_{\text{Latching}}(k)$ denotes the probability of the triple constraints (logical, electrical, and timing masking) being satisfied by a SET (with any width and amplitude) in VPO k, through the propagation path to at least one of the circuit POs and latching as a soft error at the circuit flip-flops. The triple constraints which are necessary to be satisfied by a SET to be latched as a soft error are introduced as the probability vulnerability window (PWV) [8]. Using PVW to calculate $P_{\text{Latching}}(k)$ prevents recomputing circuit SER and circuit gate vulnerability during the proposed gate sizing technique.

In the proposed technique, after resizing each gate, the probability of reaching transient faults to the VPOs of the sub-circuit in which the resized gate is located is calculated. Then, the probability of SET propagation to the circuit POs and latching as a soft error is estimated using the PVW values of the sub-circuit VPO nodes. For example, Fig. 6.4 shows the error latching probability computing procedure for gate 19 in sub-circuit 5 ($\text{ELP}_{\text{SC}_5}^{19}$). Figure 6.4a, b shows the two steps of $\text{ELP}_{\text{SC}_5}^{19}$ calculation:

Step 1: Transient fault is injected to gate 19 and the probability of transient fault propagating from gate 19 to the VPOs of sub-circuit 5 is calculated.
Step 2: The probability of the transient fault propagating from sub-circuit 5 VPOs to the circuit POs is calculated.

In the first step, the probability of the transient fault reaching to VPOs 10 and 14 is estimated by calculation of $P_{\text{gen}}(19)$, $P_{\text{prop}}(19,14)$, and $P_{\text{prop}}(19,10)$. In the next step, in order to compute the probability of the transient fault propagation to the circuit POs and latching at the flip-flops, we use the values of PVW at nodes 10 and 14. Figure 6.4b shows the PVW calculation procedure for these nodes. PVW of these nodes is calculated by circuit backward propagation from the circuit POs, as will be described in detail in Sect. 6.1.3.3.

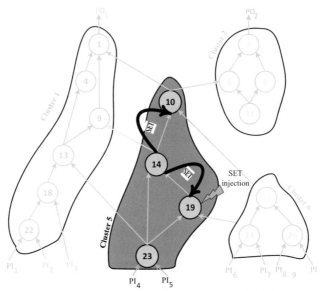

(a) Computing P_prop (19,10) ,P_prop (19,14)

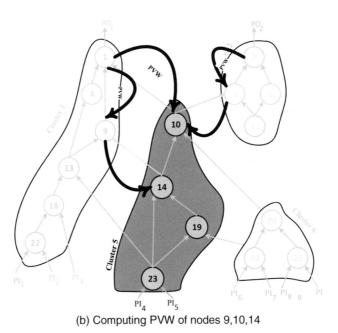

(b) Computing PVW of nodes 9,10,14

Fig. 6.4 The error probability computation procedure of node 19 in sub-circuit. (**a**) Computing $P\_prop(19, 10)$, $P\_prop(19, 14)$. (**b**) Computing PVW of nodes 9, 10, 14

It is notable that by only one backward traversing the circuit graph, the PVWs for all circuit nodes are calculated and it is not needed to recompute during different ELP calculations.

6.1.3   Gate Resizing

In the proposed technique, in order to consider the effects of gate sizing on the circuit reliability, we investigate the gate sizing effects on the sub-circuit reliability in which the resized gate is located. In fact, the key idea of the proposed technique is to evaluate the effects of gate sizing locally (sub-circuit evaluation) instead of the global evaluations (circuit evaluation) as this local effect evaluation is a perspective of global effect assessment.

The SER of the combinational circuit ($SER_{circuit}$) can be calculated by summing up the soft error rate of the individual gates (SER_i):

$$SER_{circuit} = \sum_{i=1}^{N_{node}} SER_i \tag{6.9}$$

where N_{node} is the number of gates in the circuit which may be hit by the high energetic particle. SER_i is formulated as

$$SER_i = P_{gen}(i) * P_{soft\ error}(i) \tag{6.10}$$

where $P_{gen}(i)$ is the logical probability of generating a transient fault in gate i and $P_{soft\ error}(i)$ shows the probability that the generated transient fault in gate i results in a soft error in any flip-flops at the circuit outputs. $P_{soft\ error}(i)$ is dependent on the triple masking mechanisms (i.e., logical, electrical, and timing masking) and is calculated as follows:

$$P_{soft\ error}(i) = \sum_{j=1}^{N_{ff}} P_{logic}(i,j) \cdot P_{elec}(i,j) \cdot P_{Timing}(j) \tag{6.11}$$

where N_{ff} is the number of flip-flops in the circuit output, $P_{logic}(i,j)$ shows the sum of logical probability of transient fault propagating through the path between gate i and output j, and $P_{elec}(i,j)$ is the electrical probability representing the electrical masking of the gates located in the propagation path of transient fault. $P_{Timing}(j)$ represents the error latching probability (PL) for the flip-flop which is connected to the PO j.

According to Eqs. (6.1)–(6.8), the soft error rate of a combinational circuit ($SER_{Circuit}$) could be calculated by summing up the sub-circuit error probability of the extracted sub-circuit from original circuit:

$$\text{SER}_{\text{Circuit}} = \sum_{\forall \text{Sub circuit SC}_i \in \text{Circuit}} \text{SEP}_{\text{SC}_i} \tag{6.12}$$

where SEP_{SC_i} is the error probability of sub-circuit SC_i.

Therefore, we can evaluate the effects of each gate resizing on the circuit SER by computing the SEP of a sub-circuit in which the resized gate is located.

The proposed gate sizing technique proceeds by choosing a sub-circuit and resizing its gates. Sub-circuits gate sizing is done in such a way that SEPs are reduced. By reducing SEPs of each sub-circuit, the total circuit SER will be reduced too.

6.1.3.1 Computing the Sub-circuit Error Probability

Without loss of generality, the analytical approach proposed in [9] is used in order to calculate $P_{\text{gen}}(j)$ and $P_{\text{prop}}(j,k)$ (other similar techniques can be adapted to the proposed flow). In this method, the value of $P_{\text{prop}}(j,k)$ is calculated considering the impacts of electrical and logical masking; a four-value logic and probability system is proposed to propagate transient faults. The proposed model in [10] is used to model the electrical attenuation of the transient faults.

As stated in previous section, we use PVW concept in order to compute $P_{\text{Latching}}(k)$. The PVW of the circuit nodes (i.e., gate outputs) indicates the constraints that a SET has to satisfy in order to be latched in a memory element as a soft error. A PVW can be considered as four ordered parameters which are characterized as follows [9]:

PVW · OID_w: Output ID, the memory element in which the soft errors will be latched due to a SET occurring in a given gate output.

PVW · ST_w: Starting time, the time instance which shows the starting point of the time interval of PVW.

PVW·W_w^x: Width pairs (W_w^1 and W_w^0), the pair of width values related to the time interval of PVW.

PVW · P_{Sen}: Sensitization probability, a probabilistic value which is associated with PVW.

PVW_i^j represents the vulnerability of PO j considering SETs originating in circuit node i. The set of vulnerabilities of all POs regarding node i can be defined as

$$\text{PVW}_i = \left\{ \text{PVW}_i^j \mid j \in 1 \ldots N_{\text{PO}} \right\}$$

where N_{PO} shows the number of POs.

In order to evaluate the soft error vulnerability of the circuit gate outputs, a metric called as triple constraint satisfaction probability (TCS) is defined and calculated for each gate of the circuit [8]. TCS represents the probability that a SET in the output of a given gate will satisfy the triple (logical, electrical, and timing) constraints through its path to at least one of the POs in the circuit. TCS metric for a gate G_k, TCS_k, is calculated as follows:

$$\text{TCS}_k = \text{LCS}_k * \text{ETCS}_k \tag{6.13}$$

where LCS_k is the probability that a SET at the output of G_k satisfies the logical constraint and ETCS_k shows the probability of satisfying both electrical and timing constraints by a SET at G_k in order to be latched as a soft error in the circuit.

ETCS_k and LCS_k can be calculated according to the gate PVW parameters [8]. Hence, Eq. (6.7) can be rewritten as follows:

$$\text{ELP}_{\text{SC}_i}^j = \sum_{k \in \text{VPO}_{\text{SC}_i}} P_{\text{gen}}(j) * P_{\text{prop}}(j,k) * \text{TCS}_k \tag{6.14}$$

TCS and PVW for all circuit gates are computed through the following steps:

(a) *Levelizing:* The circuit nodes in the CG of each PO are levelized using the topological sorting algorithm. Please note that the traditional topological sorting is performed with a modification on the edge direction of the circuit graph; that is, if there is an edge (u, v) in the circuit graph, then u appears after v in the list.
(b) *PVW and TCS computation*: For the nodes which are directly connected to POs, the PVW set contains one PVW which is determined by the initialization step using Table 6.1. After calculating TCS for the nodes connected to POs, the PVW set of the fan-in gates of the circuit nodes (i.e., the gates which are connected to the gate input) is computed using the computation model presented in [8]. The cycle of TCS and PVW computation procedure is repeated level by level until all gates in all levels are visited.

6.1.3.2   Sub-circuit Levelizing

As discussed in the previous sections, the circuit gate PVW parameters are dependent on their fan-out gate PVW parameters. Therefore, by changing the PVW parameters of a gate, the PVW of all gates located in its fan-in cones will be changed. On the other hand, by resizing a gate and varying its electrical property, the PVW parameters of the gates will be changed. Therefore, after resizing each gate, the PVW of the gates in its fan-in cone in the circuit should be recomputed. Recomputing the PVWs in each round of resizing optimization procedure increases the optimization algorithm runtime significantly. In the proposed technique, we cope with this issue by levelizing and updating the sub-circuits. After partitioning the circuit into a

Table 6.1 PVW parameter value of PO nodes

Parameter	Initiation value
OID_w	j
ST_w	$T_{\text{clk}} - t_{\text{s}}$
W_w^x	$t_{\text{h}} + t_{\text{s}}$

T_{clk}, t_{s}, and t_{h} are clock period, setup time, and hold time of flip-flops, respectively

set of sub-circuits, the extracted sub-circuits are levelized. Algorithm 6.1 shows the overall levelizing procedure. The algorithm works based on backward traversing of the circuit graph from POs. During backward traversing from each POs, level counter (LC) is initialized to zero and when traversing the CG, after visiting each cluster root node (cluster roots have been determined during circuit partitioning), the LC is added by one. Then, if the LC is more than the $CROL_{CL_i}$ (where $CROL_{CL_i}$ is defined as the cluster root level for cluster CL_i and initialized to zero for all clusters), LC will be assigned to the $CROL_{CL_i}$ as the new $CROL_{CL_i}$ for CL_i. Finally, the level of each sub-circuit will be equal to its CROL. After levelizing the sub-circuits, they are sorted according to their level such that the sub-circuits with the lower level are placed before another sub-circuit in the list. Figure 6.5 shows an example of the levelized sub-circuit shown in Fig. 6.3.

6.1.3.3   PVW Updating

After levelizing, the resizing algorithm proceeds with the sub-circuits in the lowest level (level 1). During resizing the gates in a sub-circuit, the PVW of the gates in the sub-circuit which are located in the fan-in cone of the resized gate is recomputed. It is worth to notice that, after resizing a gate, it is not required to recompute the PVW of the gates located in the fan-in cone of the resized gate in the main circuit. After resizing all sub-circuits in the first level, the PVW of the sub-circuit in the next level is recomputed. This procedure continues level by level until the sub-circuits in the last level are visited.

Algorithm 6.1 Sub-circuit Levelizing
1:   **Input**: clustered circuit, CRL
2:   CRL=Cluster Root List
3 : CRL_k=Cluster Root of Cluster k
4:   $CROL_{SC_i}$ =Cluster ROot Level of cluster SC_i
5:   Initiation all CROLs to zero
6:   LC=Level Counter
7:   **For** each Node N $\in PO$
9:        Set LC to zero
10:       Traverse the circuit graph from N **Until** reach to the VPIs backwardly
11:            **IF** visited Node $v \in CRL$ **then**
12:                $LC++$
13:                **IF** $LC > CROL_v$ **then**
14:                    Set $CROL_v$ to LC
15:                **End if**
16:            **End if**
17:        **End Until**
18:   **End for**

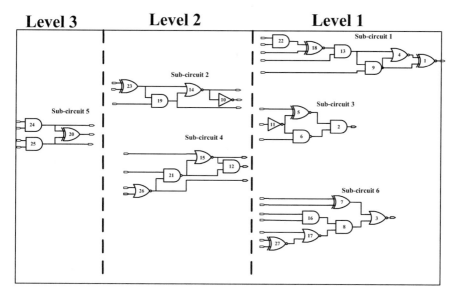

Fig. 6.5 The levelized sub-circuits

6.1.4 Sub-circuit Gate Sizing Optimization

In order to minimize the sub-circuit SER, we use an optimization method based on the genetic algorithm. It is notable that that we are able to employ other iterative optimization algorithms. In the genetic algorithm, the solution is usually encoded into a binary string called chromosome [11]. Instead of working with a single solution, the search begins with a random set of chromosomes called initial population. For a typical gate, k different sizes are considered that are presented by 0 to $k - 1$ in the chromosome, respectively, where each gene represents the size considered for one gate (Fig. 6.6).

Each chromosome is assigned a fitness that is directly related to the objective function of the optimization problem. In the proposed method, the goal of genetic algorithm is to minimize sub-circuit SEP. On the other hand, the SEP optimization is performed under a certain performance and area constraints. If the timing and area constraints are not satisfied, the configuration is not desirable and the corresponding chromosome should have little chance to survive. Therefore, the fitness function is defined as follows:

$$\text{Fitness} = \frac{1}{\text{SEP}} - \text{penalty} \qquad (6.15)$$

where the penalty is a big number if the timing and area constraints are violated, such that those chromosomes which have lower SEP and satisfy timing/area requirement have a better fitness and more chance to survive.

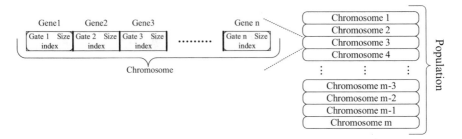

Fig. 6.6 The population and encoding scheme, where the "genes" for each gate are concatenated to form a chromosome

In order to evaluate the circuit delay, we use slack timing analysis. The slack of a gate is the amount of delay by which a gate delay may be increased without affecting the critical delay of the circuit [12]. In the proposed technique, before resizing the sub-circuit gates, the slack time of each gate in the circuit is computed to determine the circuit gates that can be downsized. During the gate sizing, we limit the downsizings to the ones that do not make gates slack negative.

The population of chromosomes is modified to a new generation by applying three operators (selection, crossover, and mutation). After iteratively applying selection, mutation, and crossover, the state of population gradually approaches the state that contains global optimal solutions. Here, we use whole arithmetic crossover [11] as the crossover operator. The inconsistent mutation operator is employed as the mutation operator [11].

6.2 Experimental Results

The proposed technique is implemented in C++ and run on a Microsoft Windows machine with a Pentium Core i7 (2.4-GHz) processor and an 8 GB RAM. All SER measurements were performed assuming a sea-level neutron flux of $56.5 \, \mathrm{m}^{-2} \, \mathrm{s}^{-1}$ [13]. We use 45 nm Nangate Open Cell Library [14] as the target library. In the genetic algorithm, the population size and maximum generation are set to 100 and 200, respectively. The crossover probability is set to 0.9 and mutation probability is set to 0.1. We explore the efficiency of the proposed optimization approach by applying the algorithm to ISCAS'85 [15] and the large-scale EPFL [16] benchmark circuits.

We apply the proposed technique and quantify the SER reduction ratio (between the baseline SER and the optimized SER) and relative circuit area and delay penalty. Table 6.2 reports the obtained results for SER reduction and area/delay overhead. In the first step of the proposed technique, the slack time of each gate has been computed, with the delay constraint set to the length of the longest structural path in the circuit. The number of the gates to be considered for resizing at the end of this step (gates with a nonzero slack time) is reported for each circuit in column 4 of Table 6.2.

Table 6.2 SER reduction, delay, and area penalty of performing the proposed technique

Benchmark information			Proposed optimization algorithm			
Circuit name	# of gates (PI, PO)	Levels	# Resizable gates	SER reduction (%)	Area change (%)	Delay change (%)
C432	160(36, 7)	30	155	7.67	9.42	0
C499	650(41, 32)	28	453	41.83	16.98	−16.37
C880	512(60, 26)	33	487	37.93	19.55	−16.69
C1355	653(41, 32)	30	613	47.62	17.84	−21.47
C1908	699(33, 25)	39	690	51.68	18.43	−7.28
C2670	756(233, 140)	38	725	49.02	16.97	−33.12
C3540	1467(50, 22)	52	1411	39.26	17.85	−18.47
C5315	2115(178, 123)	41	2057	51.69	19.97	−8.59
C6288	4507(32, 32)	122	4436	57.71	17.90	−24.12
C7552	2534(207, 108)	60	2483	44.38	16.93	−13.89
Bar	5648(135, 128)	12	5512	44.44	19.15	−44.65
Round robin	23,233(256, 129)	87	23,005	51.36	16.75	−34.4
Square	35,564(64, 128)	250	34,979	63.81	19.49	−37.17
Log2	54,494(32, 32)	444	54,284	53.57	18.87	−28.18
Average				45.85	17.57	−21.74

We optimize the circuit SER with a 20% area overhead constraint value for different benchmark circuits. The percentage of SER reduction is reported in column 5 of Table 6.2. The results show that the proposed optimization technique can achieve up to a 45% reduction in the circuit SER with imposing only 17.57% area overhead.

Although a gate delay is increased by downsizing, the proposed technique rejects the downsizings which results in negative gate slacks and thus the critical delay of the circuit does not affect. Hence, the proposed technique not only does impose any delay overhead, but also improves circuit performance by 21% on average as well due to the upsizing actions.

The impact of area constraint on the efficiency of the proposed technique is studied for ISCAS'85 and EPFL benchmark circuits and is shown in Fig. 6.7. We apply the proposed method for different cases where the area overhead is limited to 20%, 25%, and 30% of the original circuit area. As can be seen, there is an inverse relationship between the area constraint and the circuit SER improvement. As the area constraint is decreased, more SER improvement is achieved. In Fig. 6.7 the delay changes for ISCAS'85 and EPFL circuits with various area constraints (20%, 25%, and 30%) are also presented. As it can be seen, across various area constraints, a performance improvement of 0–44% is achieved. By adjusting the threshold for area overhead in the proposed technique, we can always make a trade-off between area overhead and SER reduction.

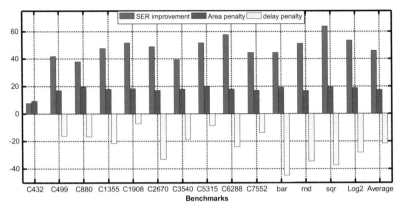

(a) SER optimization under 20% area constraint

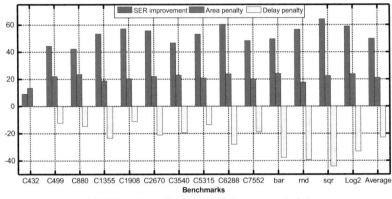

(b) SER optimization under 25% area constraint

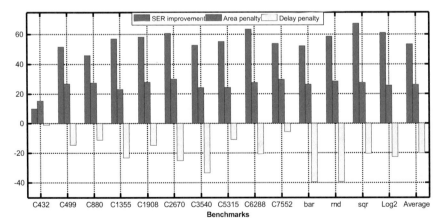

(c) SER optimization under 30% area constraint

Fig. 6.7 Impact of area constraint on SER reduction, area, and delay of benchmark circuits. (**a**) SER optimization under 20% area constraint. (**b**) SER optimization under 25% area constraint. (**c**) SER optimization under 30% area constraint

6.3 Conclusions

In this chapter, we introduce a gate sizing SER optimization technique for large-scale combinational circuits. In this technique, the circuit is split into a set of topologically levelized small sub-circuits through the cone structures. Then, the sizes of the gates in sub-circuits are optimized using a genetic optimization algorithm. Unlike the previous SER optimization techniques in which the circuit SER is used to investigate the effects of each gate sizing on circuit reliability, we introduced the sub-circuit error probability (SEP) concept to evaluate the circuit reliability during the gate sizing optimization. The results showed that such evaluation approach significantly speeds up the gate sizing optimization process.

References

1. R. R. RAO, D. BLAAUW, AND D. SYLVESTER, "SOFT ERROR REDUCTION IN COMBINATIONAL LOGIC USING GATE RESIZING AND FLIPFLOP SELECTION," IN PROC. IEEE/ACM INTERNATIONAL CONFERENCE ON COMPUTER-AIDED DESIGN (ICCAD), 2006, PP. 502–509.
2. W. SOOTKANEUNG AND K. K. SALUJA, "SOFT ERROR REDUCTION THROUGH GATE INPUT DEPENDENT WEIGHTED SIZING IN COMBINATIONAL CIRCUITS," IN PROC. 12TH INTERNATIONAL SYMPOSIUM QUALITY ELECTRONIC DESIGN (ISQED), 2011, PP. 1–8.
3. Q. ZHOU AND K. MOHANRAM, "GATE SIZING TO RADIATION HARDEN COMBINATIONAL LOGIC," IEEE TRANS. COMPUT. DES. INTEGR. CIRCUITS SYST., VOL. 25, NO. 1, PP. 155–166, 2006.
4. F. DABIRI, A. NAHAPETIAN, T. MASSEY, M. POTKONJAK, AND M. SARRAFZADEH, "GENERAL METHODOLOGY FOR SOFT-ERROR-AWARE POWER OPTIMIZATION USING GATE SIZING," IEEE TRANS. COMPUT. DES. INTEGR. CIRCUITS SYST., VOL. 27, NO. 10, PP. 1788–1797, 2008.
5. W. SHENG, "SOFT ERROR OPTIMIZATION OF STANDARD CELL CIRCUITS BASED ON GATE SIZING AND MULTI-OBJECTIVE GENETIC ALGORITHM," IN PROC. 46TH ANNUAL DESIGN AUTOMATION CONFERENCE (DAC), 2009, PP. 502–507.
6. K. BHATTACHARYA AND N. RANGANATHAN, "A UNIFIED GATE SIZING FORMULATION FOR OPTIMIZING SOFT ERROR RATE, CROSS-TALK NOISE AND POWER UNDER PROCESS VARIATIONS," IN PROC. 10TH INTERNATIONAL SYMPOSIUM QUALITY ELECTRONIC DESIGN (ISQED), 2009, PP. 388–393.
7. M. RAJI AND B. GHAVAMI, "SOFT ERROR RATE REDUCTION OF COMBINATIONAL CIRCUITS USING GATE SIZING IN THE PRESENCE OF PROCESS VARIATIONS," IEEE TRANS. VERY LARGE SCALE INTEGR. (VLSI) SYST., VOL. 24, NO. 99, PP. 1–14, 2016.
8. M. RAJI, H. PEDRAM, AND B. GHAVAMI, "A PRACTICAL METRIC FOR SOFT ERROR VULNERABILITY ANALYSIS OF COMBINATIONAL CIRCUITS," MICROELECTRON. RELIAB., VOL. 55, NO. 2, PP. 448–460, 2015.
9. M. FAZELI, S. G. MIREMADI, H. ASADI, AND S. N. AHMADIAN, "A FAST AND ACCURATE MULTI-CYCLE SOFT ERROR RATE ESTIMATION APPROACH TO RESILIENT EMBEDDED SYSTEMS DESIGN," IN PROC. INTERNATIONAL CONFERENCE ON DEPENDABLE SYSTEMS & NETWORKS (DSN), 2010, PP. 131–140.
10. R. RAJARAMANT, J. S. KIM, N. VIJAYKRISHNAN, Y. XIE, AND M. J. IRWIN, "SEAT-LA: A SOFT ERROR ANALYSIS TOOL FOR COMBINATIONAL LOGIC," IN PROC. 19TH INTERNATIONAL CONFERENCE ON VLSI DESIGN, 2006.
11. A. E. EIBEN AND J. E. SMITH, INTRODUCTION TO EVOLUTIONARY COMPUTING, VOL. 53. SPRINGER, 2003.

12. J. VYGEN, "SLACK IN STATIC TIMING ANALYSIS," IEEE TRANS. COMPUT. DES. INTEGR. CIRCUITS SYST., VOL. 25, NO. 9, PP. 1876–1885, 2006.
13. V. FERLET-CAVROIS, L. W. MASSENGILL, AND P. GOUKER, "SINGLE EVENT TRANSIENTS IN DIGITAL CMOS—A REVIEW," IEEE TRANS. NUCL. SCI., VOL. 60, NO. 3, PP. 1767–1790, 2013.
14. www.nangate.com.
15. F. BRGLEZ, "A NEUTRAL NETLIST OF 10 COMBINATIONAL BENCHMARK CIRCUITS AND A TARGET TRANSLATION IN FORTRAN," IN ISCAS-85, 1985.
16. L. AMARÚ, P.-E. GAILLARDON, AND G. DE MICHELI, "THE EPFL COMBINATIONAL BENCHMARK SUITE," IN PROC. 24TH INTERNATIONAL WORKSHOP ON LOGIC & SYNTHESIS (IWLS), 2015, NO. EPFL-CONF-207551.

Chapter 7
Resynthesize Technique for Soft Error-Tolerant Design of Combinational Circuits

Logical, electrical, and timing masking mechanisms are affective factors on the propagation of the single-event transients (SETs) in the combinational circuits. These three mechanisms avoid some SETs from affecting circuit reliability. However, the effectiveness of electrical and latching window masking properties is limited by continuing scaling trends.

In logical masking-based techniques, a combinational circuit is either restructured or resynthesized to maximize the logical masking properties of a circuit. In this chapter, we introduce a novel resynthesis soft error hardening technique for combinational circuits that is based on the partitioning, partial logical restructuring, and local reliability evaluation. In the introduced technique, the main circuit is partitioned into a set of small sub-circuits based on the minimum connection objective. Then, different logical implementations of sub-circuits with the same function as the original sub-circuit are carried out using AND-INV Graph (AIG) scheme. Among the obtained implementations of the sub-circuit, an implementation that provides most circuit SER reduction and meanwhile does not violate the original circuit time and area constraints is replaced instead of the current sub-circuit in the circuit. This procedure continues for all extracted sub-circuits. In order to evaluate the effects of manipulating a sub-circuit on circuit reliability, we introduce a global failure probability (GFP) metric which provides an estimation of the contribution of a sub-circuit in total circuit SER. We use GFP to evaluate the impacts of replacing different implementations of each sub-circuit on circuit SER improvement, instead of recomputing the total circuit SER in each round of resynthesis process. Such a reliability evaluation technique significantly reduces the resynthesis process runtime. The main innovation of this chapter is to introduce a new parameter to evaluate circuit SER after resynthesis without needing complex calculations.

The rest of this chapter is organized as follows: Section 7.1 provides a motivation. In Sect. 7.2, we present the proposed method for SER reduction. Section 7.3 shows the experimental results. Finally, the conclusions are provided in Sect. 7.4.

© Springer Nature Switzerland AG 2021
B. Ghavami, M. Raji, *Soft Error Reliability of VLSI Circuits*,
https://doi.org/10.1007/978-3-030-51610-9_7

7.1   Soft Error in Combinational Circuits

Figure 7.1 shows a general view of a combinational circuit including the combinational logic gates, as well as primary inputs (PI) and primary outputs (PO). A SET may appear in the output of the internal logic gates in addition to the interconnection between PIs and internal gates. The SET originated at any gates may be latched in each element considered in POs.

The soft error rate of a combinational circuit C ($SER_{Circuit}$) can be calculated by summing up the soft error rate of individual gates (SER_{G_i}) in the circuit [1, 2]:

$$SER_{Circuit} = \sum_{i=1}^{N} SER_{G_i} \qquad (7.1)$$

where N is the number of gates in the circuit which may be hit by the high energetic particle. Each SER_{G_i} can be further formulated by integrating over the range $q = 0$ to Q_{MAX} the products of particle hit rate and the probability that a soft error can survive. Therefore,

$$SER_{G_i} = \int_{q=0}^{Q_{MAX}} \left(R_q * P_{soft\ error} \left(G_i, q \right) \right) dq \qquad (7.2)$$

where $P_{soft\ error}(i, q)$ represents the probability that a transient fault originated from the particle of charge q at gate G_i can result in one soft error at any flip-flop. Also, R_q represents the effective frequency for a particle hit of charge q in unit time according to [1]. That is,

$$R_q = F * K * A * \frac{1}{Q_s} * \exp\left(\frac{-q}{Q_s} \right) \qquad (7.3)$$

where F, K, A, and Q_s denote neutron flux (>10 MeV), technology-independent fitting parameter, susceptible area in cm², and charge collection slop, respectively. For

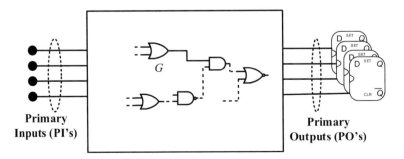

Fig. 7.1 A general view of a combinational circuit

a practical SER analysis, the continuous integration in Eq. (7.2) is often approximated by a sum of discretized charges [2]. That is,

$$\text{SER}_{G_i} = \sum_{K=1}^{n} R_{q_K} * P_{\text{soft error}}\left(G_i, q_K\right) \tag{7.4}$$

where $q_K = K * (q_{\max} - q_{\min})/n$. Also, $P_{\text{soft error}}(G_i, q_K)$ is calculated as follows [3]:

$$P_{\text{soft error}}\left(G_i, q_K\right) = 1 - \prod_{j \in Os}\left(1 - \text{LP}_{O_j}\left(G_i, q_K\right)\right) \tag{7.5}$$

where O is the number of flip-flops connected to the circuit outputs. Also, $\text{LP}_{O_j}\left(G_i, q_K\right)$ is the latching probability of the propagated waveform caused by transient fault at the gate G_i at flip-flop connected to output O_j. $\text{LP}_{O_j}\left(G_i, q_K\right)$ is calculated as follows:

$$\text{LP}_{O_j}\left(G_i, q_K\right) = \frac{T_h + E_{O_j}^{\text{PW}}\left(G_i, q_K\right) + T_s}{T_{\text{clK}}} \tag{7.6}$$

where T_s, T_h, and T_{clk} are setup time, hold time, and clock period of the flip-flop connected to circuit output O_j. Also, $E_{O_j}^{\text{PW}}\left(G_i, q_K\right)$ is the expected pulse width of the propagated waveform caused by transient fault to a primary output O_j. $E_{O_j}^{\text{PW}}\left(G_i, q_K\right)$ is calculated as follows:

$$E_{O_j}^{\text{PW}}\left(G_i, q_K\right) = \sum_{K \in \text{Event List of } O_j}\left(P_{0^e}^K + P_{1^e}^K\right)\left(T_{K+1} - T_K\right) \tag{7.7}$$

where $P_{0^e}^k$ ($P_{1^e}^k$) is the probability that event k has the value of 0^e (1^e) and T_K is the time of kth event of the event list.

In general, evaluating the effect of restructuring either the whole or a small part of the circuit on its reliability needs recomputing of the total circuit SER. On the one hand, SER computation is very time consuming for middle- and large-size circuits, and on the other hand any reliability enhancement technique needs to compute circuit SER repeatedly during its improvement process consequently, and previous reliability evaluation techniques which are all based on repeatedly computing circuit SER cannot be used in practice.

To address this issue, we introduce a soft error-tolerant circuit design method in which the circuit is split into a set of small sub-circuits, both single output and multi-output sub-circuits. The proposed method helps circuit partitioning in two aspects. First, since the entire circuit is divided into small sub-circuits, we are able to manipulate all susceptible parts of the circuit to soft errors, while the previous methods are suffering from the serious shortcoming as they are not capable of manipulating all circuit parts. The second and more important aspect of the proposed circuit partitioning is that the effect of any logical manipulation of a small

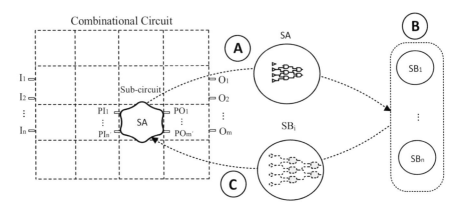

Fig. 7.2 The procedure of manipulation and substitution of a sub-circuit in a combinational circuit

part of the circuit on the total circuit reliability is studied locally using a new metric called global failure probability (GFP). GFP, which is an estimation of sub-circuit contribution on total circuit SER, considers sub-circuit structure and location as well as sub-circuit roles on the transient fault generation and propagation in the main circuit. It is shown that manipulating and reducing sub-circuit GFP result in reducing of the total circuit SER.

Figure 7.2 shows the procedure of manipulation and substitution of a sub-circuit in a combinational circuit containing n primary inputs ($I_1, I_2, \ldots, I_n$) and m primary outputs ($O_1, O_2, \ldots, O_n$). This procedure consists of three main steps. In step (A), sub-circuit SA is chosen among all extracted sub-circuits. In the next step (B), different logical implementations (SB$_1$, SB$_2$, \ldots, SB$_n$) of sub-circuit SA which do not violate predetermined circuit constraints are carried out and their corresponding GFP is calculated. Finally, in step (C), the implementation of SA with the least GFP (SB$_i$) is replaced by SA in the main circuit.

7.2   The Proposed Soft Error-Tolerant Method

Figure 7.3 shows the overall flow of the proposed method in order to reduce combinational circuit SER. As shown in the figure, at first, the main circuit is partitioned into a set of small sub-circuits (SC-Set). In the next step, for each sub-circuit SC$_i$, different logical implementations of SC$_i$ are extracted. Then, the best implementation of SC$_i$ is chosen using GFP parameter and is replaced in the circuit. This process continues until the area constraint is violated. In the following, we describe the proposed method in detail.

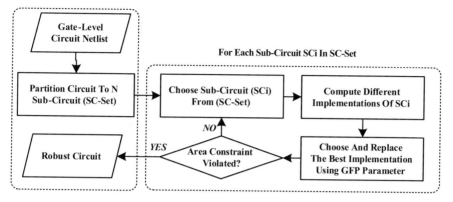

Fig. 7.3 The proposed method based on partitioning and local logical replacement

7.2.1   Circuit Partitioning

Nowadays, advances in technology lead to increase of the design complexity of the VLSI digital circuits. A common strategy is to partition the design into smaller portions, while each of which can be processed with some degree of independence. In this paper, as the first step of the proposed SER reduction method, the circuit is partitioned into some sub-circuits so that the computation complexity is reduced as each sub-circuit is processed independently.

In order to partition the digital circuits, the circuit is considered as a graph $G(V, E)$, while the nodes of the graph ($V = \{v1, v2, ..., vn\}$) represent gates of the circuit and the edges ($E = \{e1, e2, ..., en\}$) represent the connections between the gates. Therefore, it is possible to partition a combinational circuit with n logical gates ($|V| = n$) into k small sub-circuits (SC^K-Set = $\{SC_1, SC_2, ..., SC_K\}$); while each sub-circuit is a subset of the main circuit (7.8), the sub-circuits have no intersection with each other (7.9) and the union of all k sub-circuits is equal to the main circuit (7.10). Also, an edge between two different partitions is called a *cut* or *connection*:

$$SC_i \in C \quad i = \{1,2,...,k\} \tag{7.8}$$

$$SC_i \cap SC_j = \varnothing \quad i \neq j \quad i,j = \{1,2,...,k\} \tag{7.9}$$

$$\bigcup_{i=1}^{k}(SC_i) = C \tag{7.10}$$

Partitioning a combinational circuit C into a set of sub-circuits is such a way that each partition meets the cost constraints. The traditional cost constraints are the number of sub-circuits, sub-circuit size, and number of connections between sub-circuits which is defined as the cut size. As mentioned before, partitioning a circuit into some smaller sub-circuit leads to some degrees of independence in processing them. If a sub-circuit processes without considering other sub-circuits, the connection between sub-circuits may affect the efficiency of the proposed method and thus may result in decreasing the reliability of the circuit. Furthermore, the large number of connections between sub-circuits may lead to dependencies between them.

Since the number of connections (or the cut size) plays an important role in the propagation of transient faults through the sub-circuits to the primary output of the circuit, we only consider the number of connections between sub-circuits as a constraint in partitioning the combinational circuit. Therefore, the problem here is partitioning the circuit into some sub-circuits with the goal of minimizing the number of connections between sub-circuits. The minimum connection (cut size) objective, obj, is described as follows:

$$\text{obj} = \sum_{i=1}^{k}\sum_{j=1}^{k} C_{ij} \quad i \neq j \quad i, j = \{1, 2, \ldots, k\} \tag{7.11}$$

where C_{ij} shows the number of connections between sub-circuits SC_i and SC_j and k is the number of sub-circuits obtained from partitioning of the main circuit.

7.2.2   Local Sub-circuit Evaluation

For each sub-circuit, the *fan-in cone* and the *fan-out cone* are defined as follows:

- *Fan-in cone* of the sub-circuit is the gates of the circuit from which there is a path to inputs of the sub-circuit.
- *Fan-out cone* of the sub-circuit is the gates of the circuit to which there is a path from outputs of the sub-circuit.

The fan-in cone and fan-out cone of a sub-circuit are shown in Fig. 7.4. The sizes of the fan-in cone and fan-out cone of the sub-circuits are different according to their location in the main circuit.

By considering the fan-in cone (FIC) and fan-out cone (FOC) of the sub-circuit, the role of the sub-circuits in circuit soft errors includes the transient fault generation and the transient fault propagation. The sub-circuit role in transient fault generation refers to the case where a transient fault is generated at the gate g of the sub-circuit and this fault propagates through the FOC of the sub-circuit and arrives at the primary outputs (POs). Also, the sub-circuit role in transient fault propagation refers to the case where a transient fault is generated at gate g in the FIC of the sub-circuit and this fault propagates through the sub-circuit until it reaches the POs.

As an example, the process of transient fault generation in a sub-circuit is shown in Fig. 7.4. As shown in this figure, the transient fault is generated in gate Gg of the sub-circuit and after propagating to sub-circuit outputs, it reached the POs. Also, the procedure of the transient fault propagation through the sub-circuit is shown in this figure. As can be seen in this figure, the transient fault is generated at gate Gp in the

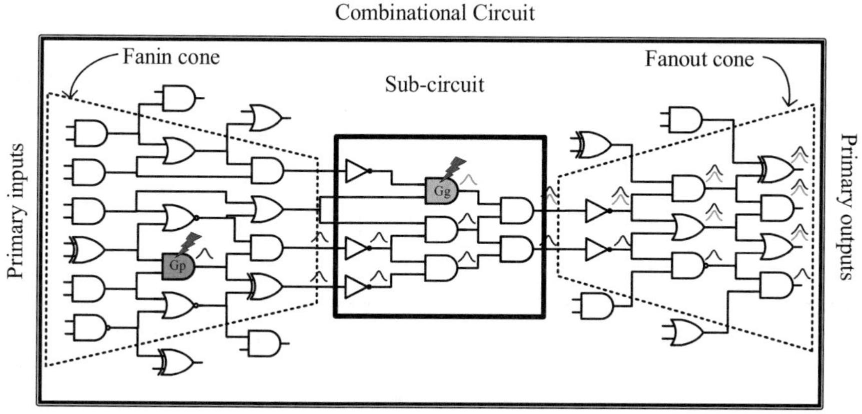

Combinational Circuit

Fig. 7.4 FIC and FOC of a sub-circuit. The transient fault generation by the sub-circuit and the transient fault propagation through the sub-circuit

FIC of the sub-circuit and after propagating through the sub-circuit it reached the POs.

In order to evaluate the impact of a sub-circuit on both transient fault generation and propagation globally, global failure probability (GFP) metric which includes both generation and propagation effects is introduced as follows:

$$\text{GFP}_{SC_i} = P_{\text{generation}} + P_{\text{propagation}} \tag{7.12}$$

where $P_{\text{generation}}$ shows the probability that a transient fault is generated in the sub-circuit and results in a soft error and $P_{\text{propagation}}$ is the probability that a transient fault is propagated through the sub-circuit and results in a soft error. The probability of a soft error occurrence due to the transient fault generation in the sub-circuit ($P_{\text{generation}}$) is equal to the sum of the probability that a soft error is originated by transient fault generation at each gate of the sub-circuit which will reach the POs. Therefore, the probability of transient fault generation by the sub-circuit ($P_{\text{generation}}$) is computed as follows:

$$P_{\text{generation}} = \sum_{i=1}^{N(\text{PG})} \text{LEG}(G_i) \tag{7.13}$$

where $\text{LEG}(G_i)$ is the probability of a soft error occurrence due to a transient fault generation at gate G_i in the sub-circuit which reaches the POs and $N(\text{PG})$ is the number of gates in the sub-circuit.

Considering Fig. 7.4, the process of changing a transient fault generated at gate G_i in the sub-circuit into a soft error consists of two steps:

The transient fault is generated at gate G_i in the sub-circuit and propagated to the sub-circuit outputs. The transient fault is propagated from the sub-circuit outputs to the circuit POs.

The probability values in step 1 are dependent on the structure of the sub-circuit while the probability values in step 2 depend on the structure of its FOC. Since restructuring the sub-circuit only changes the probability values in step 1, step 1 and step 2 can be considered separately. Hence, the probability of observing a soft error due to a transient fault generation at gate G_i in the sub-circuit is obtained as follows:

$$LEG(G_i) = \sum_{K=1}^{N(PO)} EPP_{G_i}(PO_K) \times SERG(PO_K) \tag{7.14}$$

where $N(PO)$ is the number of the sub-circuit outputs, $EPP_{G_i}(PO_K)$ shows the probability of propagating the transient fault which has been generated at gate G_i to output PO_K of the sub-circuit (step 1), $SERG(PO_K)$ is the probability that the transient fault which has been observed at output PO_K of the sub-circuit is latched to at least a flip-flop at POs of the circuit.

In order to compute the probability that the transient fault generated at gate G_i in the sub-circuit is propagated to the sub-circuit outputs, the error propagation probability rules proposed in [3] are used (Fig. 7.5b). Finally, the probability of latching the transient fault at the flip-flop connected to the output PO_K by considering logical masking is computed as

$$EPP_{G_i}(PO_K) = \frac{T_h + E_{PO_K}^{PW} + T_s}{T_{clk}} \tag{7.15}$$

In order to arrive at the POs, the transient fault has to propagate through the FOC of the sub-circuit. The FOC of the sub-circuit is the union of all FOCs of the sub-circuit outputs. The FOC of the sub-circuit output PO_1 is shown in Fig. 7.5c. Since the FOCs of the sub-circuit outputs have different structures, they have different impacts on fault propagation. In order to consider the impact of each sub-circuit output on the propagation of the transient fault to POs, $SERG(PO_K)$ is defined and computed for each sub-circuit output as follows:

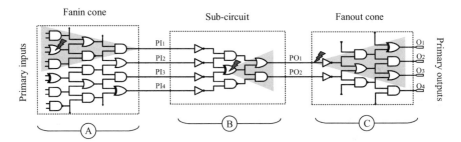

Fig. 7.5 (a) FIC of the sub-circuit input PI_j and computing SERP(PI_1). (b) Transient fault generation at the gate of the sub-circuit and computing EPPG$_i$(PO$_K$). (c) FOC of the sub-circuit input PI_j and computing SERG(PO$_1$)

$$\text{SERG}\left(\text{PO}_K\right) = 1 - \prod_{j \in \text{NFO}(\text{PO}_K)} \left(1 - \text{LP}_{O_j}\left(\text{PO}_K\right)\right) \tag{7.16}$$

where $\text{NFO}(\text{PO}_K)$ is the number of POs existing in the FOC of the sub-circuit output K, and $\text{LP}_{O_j}\left(\text{PO}_K\right)$ shows the probability of latching the transient fault at the flip-flop connected to the primary output j that exists in FOC of the sub-circuit output k (PO_K).

The probability of transient fault propagation through the sub-circuit is the probability of transient fault generation of all gates of the sub-circuit FIC, transient fault propagation through the sub-circuit, and transient fault latching at the flip-flops connected to POs. Therefore, the probability of the transient fault propagation through the sub-circuit is defined as follows:

$$P_{\text{propagation}} = \sum_{j=1}^{N(\text{PI})} \text{LEP}\left(\text{PI}_j\right) \tag{7.17}$$

where $\text{LEP}(\text{PI}_j)$ is the probability of transient fault generation in the FIC of the sub-circuit input PI_j, fault propagation through the sub-circuit, and fault latching at the flip-flops at POs.

The process of changing a transient fault generated at gate G_i in FIC of the sub-circuit into a soft error consists of three steps:

- Transient fault generation in FIC of the sub-circuit and propagation to the sub-circuit inputs
- Transient fault propagation from the sub-circuit inputs to the sub-circuit outputs
- Transient fault propagation from the sub-circuit outputs to POs

The probability value of the event in step 2 depends on the structure of the sub-circuit while the probability values of what happened in steps 1 and 3 are, respectively, dependent on the structure of the sub-circuit FIC and FOC. Since the sub-circuit restructuring only changes the probability value of the event in step 2, these three steps can be considered separately.

The probability of a soft error due to transient fault generation in FIC of sub-circuit input PI_j and then propagating through the sub-circuit and reaching at a PO is computed as follows:

$$\text{LEP}\left(\text{PI}_j\right) = \text{SERP}\left(\text{PI}_j\right) \times \text{LEG}\left(\text{PI}_j\right) \tag{7.18}$$

where $\text{SERP}(\text{PI}_j)$ is the probability of transient fault generation in FIC of the sub-circuit input PI_j (step 1) and $\text{LEG}(\text{PI}_j)$ shows the probability of transient fault generation at the sub-circuit input PI_j and reaching the POs (step 2 and step 3).

A transient fault is propagated through the FIC of a given sub-circuit before it reaches the sub-circuit inputs. FIC of a sub-circuit is the union of all FICs of the sub-circuit inputs. As an example, FIC of a sub-circuit input PI_1 is shown in Fig. 7.5a.

Since FICs of the sub-circuit inputs have different structures, they have different impacts on transient fault propagation to the sub-circuit inputs. In order to consider the contribution of each sub-circuit input in propagating the transient fault, SERP(PI$_j$) computed for each sub-circuit output is as follows:

$$\text{SERP}\left(\text{PI}_j\right) = \sum_{h=1}^{\text{NFI}\left(\text{PI}_j\right)} \text{EPP}_{G_h}\left(\text{PI}_j\right) \tag{7.19}$$

where NFI(PI$_j$) is the number of gates in FIC of the sub-circuit input j and $\text{EPP}_{G_h}\left(\text{PI}_j\right)$ shows the probability of latching a transient fault which is generated in gate G_h at flip-flop that is considered to be connected to the sub-circuit input j.

Considering Eqs. (7.12)–(7.19), GFP$_{\text{SC}_i}$ is defined as follows:

$$\text{GFP}_{\text{SC}_i} = \left[\sum_{i=1}^{N(\text{PG})}\left(\sum_{K=1}^{N(\text{PO})} \text{EPP}_{G_i}\left(\text{PO}_K\right) \times \text{SERG}\left(\text{PO}_K\right)\right)\right]$$
$$+ \left[\sum_{j=1}^{N(\text{PI})}\left(\text{SERP}\left(\text{PI}_j\right) \times \left(\sum_{K=1}^{N(\text{PO})} \text{EPP}_{\text{PI}_j}\left(\text{PO}_K\right) \times \text{SERG}\left(\text{PO}_K\right)\right)\right)\right] \tag{7.20}$$

As shown in Eq. (7.20), GFP$_{\text{SC}_i}$ depends on SERP(PI$_j$), SERG(PO$_K$), and EPP(PO$_K$). As mentioned before, SERP(PI$_j$) shows the effects of the sub-circuit FIC, SERG(PO$_K$) considers the effects of the sub-circuit FOC in transient fault propagation, and EPP(PO$_K$) indicates the effects of the transient fault generation at gates of the sub-circuit or the transient fault propagation through the sub-circuit. Therefore, since SERP(PI$_j$) and SERG(PO$_K$) are, respectively, computed according to the sub-circuit inputs and outputs, it is only enough to recompute EPP(PO$_K$) considering the sub-circuit structure (and its different implementations) again when re-evaluating the circuit SER with different implementations of the considered sub-circuit. So, by using GFP$_{\text{SC}_i}$ metric, it is possible to evaluate the effect of replacing different implementations of each sub-circuit on circuit SER improvement, instead of computing the total circuit SER.

7.2.3   Sub-circuit Restructuring

In order to increase the logical masking of a combinational circuit, the sub-circuits should be restructured while keeping their functionality in different structures. Hence, different logical implementations of the sub-circuit with equal functionality are extracted. Since sub-circuit implementations have different structures (i.e., number of gates, connection between gates and number of possible logical paths), they have different effects on generation and propagation of transient faults.

AND-INV Graph (AIG) is a method to implement the logical functions [4]. This graph only consists of AND gates and inverters (INV). An AIG is a directed acyclic graph (DAG), in which a node has either 0 or 2 incoming edges. A node with no incoming edges is a primary input (PI). A node with no outgoing edges is a primary output (PO). A node with two incoming edges is a two-input AND gate. An edge is either complemented or not. A complemented edge indicates the inversion of the signal [5]. AIG is not canonical corresponding to a logical function. In other words, it is possible to extract different implementations from a logical function using AIG method. For example, different AIG structures for logical function $F = abc$ are shown in Fig. 7.6.

As AIG method does not create canonical forms for a logical function F, it is possible to extract various implementations of F considering different design parameters (such as delay, area, number of gates, and levels). In this chapter, in order to compute different implementations of a sub-circuit, at first the graph of AIG corresponding to the sub-circuit function is created and then different implementations of the sub-circuit are extracted considering design parameters and the standard cell library. Finally, a set (API_{SC}) is formed including different implementations of the sub-circuit.

7.2.4   Optimization Algorithm

The proposed algorithm of reducing the combinational circuit SER based on partitioning and local logical replacement is shown in Algorithm 7.1. As shown in the figure, this algorithm receives a combinational circuit as an input and then computes SER and area of the circuit while values of SERG and SERP are computed for each gate at the same time. Then, the main circuit is partitioned into N small sub-circuits (SC-Set). In the next step, for each sub-circuit SC_i of the circuit, GFP_{SC_i} is computed. After that, different implementations of the sub-circuit SC_i (API_{SC_i}) are computed using AIG method. Using computed values of GFP parameter, the best

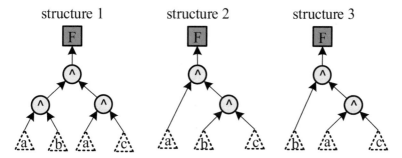

Fig. 7.6 Different AIG structures for function $F = abc$ [5]. Nodes with no incoming edges are primary inputs (a, b, and c), nodes with no outgoing edges are primary outputs (F), and nodes with two incoming edges are two-input AND gates

implementation of the sub-circuit (I_m) is chosen in the next step and replaced in the circuit while ($\text{GFP}_{I_m} < \text{GFP}_{SC_i}$). This process continues until the area constraint is violated. Note that it is possible that no solution gets found under a certain area constraint. In this case, the proposed algorithm will return the original circuit. In this case, a common solution is that the algorithm returns an improved circuit that is close to meet the area constraint.

In the ordinary circuit SER improvement using circuit restructuring (without circuit partitioning), the order of complexity is $O(D \times N^2)$ where D is different structures of the circuit and N is the number of gates in the circuit. Also, the order of complexity of the circuit SER improvement using the circuit partitioning and sub-circuit restructuring is $O(k * d * n^2)$ where k is the number of the sub-circuits in the circuit, d is the different structures of sub-circuits, and n is the number of gates in the sub-circuits. The number of gates in the circuit is more than the number of gates in the sub-circuit ($N \gg n$) and different structures of the circuit in ordinary method are too higher than the proposed method ($D \gg k * d$). Therefore, the order of complexity of the proposed method is less than that of the ordinary method.

Algorithm 7.1 Optimization Algorithm

1.   **input** = golden circuit: $\mathbf{C_G}$
2.   **output** = robust circuit: $\mathbf{C_R}$
3.   compute SER of golden circuit, SER_{CG};
4.   compute Area of golden circuit, $Area_{CG}$;
5.   *// compute SERG and SERP for all gates of circuit;*
6.   partition C_G into N Sub-Circuit and add them to SC-Set;
7.   **for each**(sub-circuit SC_i in SC-Set)
8.   *//extract SERG for all outputs of SC_i;*
9.   *//extract SERP for all inputs of SC_i;*
10.   compute GFP parameter for sub-circuit SC_i, GFP_{SCi};
11.   compute all possible implementations of SC_i using AIG, API_{SCi}; //
12.   compute GFP for all implementations in API_{SCi};
13.   Choose implementation I_m in API_{SCi} with least GFP_{Im};
14.   **if**($GFP_{Im} < GFP_{SCi}$)
15.   **swap** sub-circuit SC_i and implementation I_m;
16.   **end if**;
17.   **if**(Area constraint violated)
18.   **go to** line 21;
19.   **end if**;
20.   **end foreach**;
21.   compute SER of robust circuit SER_{CR};
22.   compute $Area$ of robust circuit $Area_{CR}$;

7.3 Experimental Results

The proposed method is implemented in C++ and run on a windows machine with a core i5 Intel processor (2.53 GHz) and 6 GB RAM. The method is applied to ISCAS'85 benchmark circuits. Also, 45 nm Nangate Open Cell Library [6] is used. All SER measurements were performed assuming a sea-level neutron flux of 56.5 m^{-2} s^{-1} [1].

In the experiments, METIS tool proposed in [7] is used for dividing the circuit into a set of small sub-circuits. METIS can partition a circuit into a user-specified number (k) of sub-circuits using the multilevel k-way partitioning paradigm. Since the number of connections between sub-circuits plays an important role in the propagation of transient faults through the sub-circuits to the primary output of the circuit, in our experiments, we only consider the number of connections between sub-circuits as a constraint in partitioning the combinational circuit. Therefore, the main goal is partitioning the circuit into some sub-circuits while minimizing the number of connections between sub-circuits. ABC tool [4] is used to extract different logical implementations of sub-circuits with the same logical function. In the experiments, different implementations of the sub-circuit are extracted using ABC tool considering design parameters and standard cell library. In this chapter, in order to extract different implementations of a sub-circuit with the same logical function, some logic synthesis scripts (such as *resyn, resyn2, resyn2rs*, and *resyn3*) introduced in [4] are used. These scripts perform some iterations of *rewrite, refactor,* and *balance* commands. By using these scripts and changing their parameters, different implementations of a sub-circuit with the same logical function are extracted. We refer the reader to [4] for further details about this tool. Also, average number of different implementations explored for each sub-circuit is about 13.

Table 7.1 shows the results of the proposed method on ISCAS'85 benchmark circuits. The first part of Table 7.1 represents the information of the benchmarks;

Table 7.1 Simulation results on ISCAS'85 benchmark circuits

Benchmark information				Proposed method		
Bench	# PI	# PO	# Gate	# SC	SER reduction (%)	Area change (%)	Delay change (%)
C432	36	7	160	8	1.98E+01	−1.85E+00	2.76E+00
C499	41	32	650	33	2.19E+01	1.29E+01	1.03E+01
C880	60	26	512	26	2.43E+01	−3.51E+00	−3.11E+00
C1355	41	33	653	33	2.75E+01	2.12E+01	7.59E+00
C1908	33	25	699	35	1.81E+01	−5.03E+00	−3.61E+00
C2670	157	64	756	38	2.26E+01	−6.15E+00	4.82E+00
C3540	50	22	1467	74	2.76E+01	1.20E+01	9.36E+00
C5315	178	123	2115	106	1.79E+01	1.58E+01	8.44E+00
C6288	32	32	4507	226	2.55E+01	2.37E+01	1.33E+01
C7552	207	108	2534	127	2.88E+01	1.82E+01	6.56E+00
Average	–	–	–	–	2.34E+01	8.72E+00	5.63E+00

column 1 shows the circuit name and the number of PIs, number of POs, number of gates, and number of sub-circuit extracted for each circuit are represented in the next columns, respectively. The second part of Table 7.1 shows the SER reduction, area overhead, and delay overhead. The proposed method tries to reduce the probability of the transient fault propagation through the circuit gates to the circuit primary outputs and increase logical masking, using sub-circuit restructuring. Also, the proposed approach tries to decrease the expected pulse width, which reaches the circuit primary output, by sub-circuit restructuring. Hence, the latching probability and the circuit SER are decreased. As shown in this table, on average, the SER of the benchmark circuits is reduced by 23.4% where the average area overhead and delay overhead are 8.72% and 5.63%, respectively. Note that, in the proposed approach, we just considered area constraint in the optimization algorithm and did not consider delay constraint. At the end of the optimization process, we measured the area overhead and delay overhead of the improved circuits.

7.4 Conclusion

In this chapter, a resynthesis technique was introduced in order to reduce the soft error rate (SER) of combinational circuits. This technique was based on the circuit partitioning and a local logical replacement. The proposed method tried to maximize the logical masking probability of the circuit gates. In this technique, the circuit was divided into a set of small sub-circuits. Then different logical implementations of sub-circuits with the same function which satisfy the circuit timing and area constraints were carried out. Then, an implementation of each sub-circuit that provided the maximum logical masking and minimum transient fault generation probability was used in place of the current sub-circuit in the main circuit. In order to choose the best alternative among the different implementations of a sub-circuit, a new metric called global failure probability (GFP) was introduced. The proposed parameter evaluated the contribution of different implementations of sub-circuits in total circuit SER. Use of GFP to assess the effect of different implementations of a sub-circuit in the circuit SER avoided estimating the total circuit SER repeatedly during SER optimization process that is very time consuming.

References

1. H.-K. Peng, C. H.-P. Wen, and J. Bhadra, "On soft error rate analysis of scaled CMOS designs: a statistical perspective," in Proceedings of the 2009 International Conference on Computer-Aided Design, 2009, pp. 157–163.
2. A. C.-C. Chang, R. H.-M. Huang, and C. H.-P. Wen, "CASSER: a closed-form analysis framework for statistical soft error rate," IEEE Trans. Very Large Scale Integr. Syst., vol. 21, no. 10, pp. 1837–1848, 2013.

3. M. Fazeli, S. N. Ahmadian, S. G. Miremadi, H. Asadi, and M. B. Tahoori, "Soft error rate estimation of digital circuits in the presence of multiple event transients (METs)," in 2011 Design, Automation & Test in Europe, 2011, pp. 1–6.

4. A. Mishchenko, "ABC: A system for sequential synthesis and verification (2007)," URL https://www.eecs.berkeley.edu/alanmi/abc, 2010.

5. A. Mishchenko, S. Chatterjee, and R. Brayton, "DAG-aware AIG rewriting: A fresh look at combinational logic synthesis," in Design Automation Conference, 2006 43rd ACM/IEEE, 2006, pp. 532–535.

6. S. Nangate, "California (2008). 45 nm open cell library," URL https://www.nangate.com, 2008.

7. G. Karypis and V. Kumar, "A fast and high quality multilevel scheme for partitioning irregular graphs," SIAM J. Sci. Comput., vol. 20, no. 1, pp. 359–392, 1998.

Index

© Springer Nature Switzerland AG 2021   109
B. Ghavami, M. Raji, *Soft Error Reliability of VLSI Circuits*,
https://doi.org/10.1007/978-3-030-51610-9

Printed in the United States
by Baker & Taylor Publisher Services